Introduction to Dyna

IAN PERCIVAL

Professor of Applied Mathematics, Queen Mary College
University of London

DEREK RICHARDS

Lecturer in Mathematics, Open University

CAMBRIDGE
UNIVERSITY PRESS

Published by the Press Syndicate of the University of Cambridge
The Pitt Building, Trumpington Street, Cambridge CB2 1RP
40 West 20th Street, New York, NY 10011-4211 USA
10 Stamford Road, Oakleigh, Melbourne 3166, Australia

First published 1982
Reprinted 1985, 1987, 1989, 1991, 1994

Library of Congress catalogue card number: 81–15514

British Library cataloguing in publication data
Percival, Ian
Introduction to dynamics.

1. Dynamics.
I. Title   II. Richards, Derek
531   QC133

ISBN 0 521 28149 0 paperback

Transferred to digital printing 1999

CONTENTS

Contents

Contents

PREFACE

Modern dynamics owes as much to Poincaré and Liapounov as to Lagrange and Hamilton, so we introduce Hamiltonian dynamics through the qualitative theory of differential equations and we highlight the geometry of phase curves and the theory of stability. Each subject, from the elementary theory of first-order systems, up to the discoveries on chaotic motion in recent decades, is introduced through simple examples. The mathematical background required of the reader is confined to 2×2 matrices, ordinary differential equations and the calculus of two variables (apart from appendix 1). Some knowledge of elementary Newtonian mechanics would be helpful, but we include other applications, including the dynamics of biological populations.

For simplicity we restrict our attention to first- and second-order systems and to Hamiltonian systems with one degree of freedom. This approach is not nearly so restrictive as one might think and enables us to introduce to undergraduates many important ideas that have previously been confined to graduate teaching or research.

The stronger connexions between the chapters are illustrated in the diagram at the end of the preface, but more advanced students or research workers should find that they can usefully dip into the book with a little cross-reference.

Very many colleagues and students have helped us. In particular we thank Brian Chirgwin, Barry Hughes, Alan Jeffrey, Mike Simpson and the referees of the Cambridge University Press for their advice and criticism of early drafts. We are grateful to Michel Hénon for lending us the originals of figures for chapter 11, to the editor of the *Quarterly of Applied Mathematics* for permission to publish them and to John Greene and Mike Lieberman for helping to clarify some of our ideas on that chapter. Like so many people, we are grateful to Joe Ford for infecting us with his enthusiasm for modern dynamics. We could not have finished the book so quickly without diligent typing assistance from Frances Thomas and assistance with the computing from Alf Vella. We could not have written it at all without the encouragement of Jill and Helen, to whom this book is dedicated.

Ian Percival	Derek Richards
London	Milton Keynes

Stronger connexions between chapters

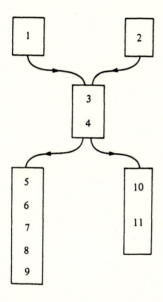

1 FIRST-ORDER AUTONOMOUS SYSTEMS

1.1 Basic theory

Dynamics is concerned with the *motion* of systems, that is, their change of state with time t.

A *first-order* system is the simplest type of dynamical system and it is defined by two properties:

(F1) The state of the system is represented by a single real variable x, which may be considered as a coordinate of a point in an abstract one-dimensional space named the *phase space*.

(F2) The motion of the system is represented by a function $x(t)$ of time satisfying a first-order differential equation

$$\frac{dx}{dt} \equiv \dot{x} = v(x, t), \tag{1.1}$$

where v is a given sufficiently well-behaved *velocity function* of x and t, whose value for a particular x and t is the *phase velocity*. The differential equation (1.1) is the *equation of motion* or *equation of change* of the system.

Radioactive decay, population changes in biological species, simple chemical reactions, the fall of a light body through a very viscous fluid and the discharge of an electrical condenser through a resistance are all examples of first-order dynamical systems. Newtonian mechanical systems are not of first order, except in extreme limiting cases like the light body in the viscous fluid, so Newtonian systems will be considered later.

For the rest of this chapter we restrict our attention to first-order *autonomous* systems, because they are particularly simple. An autonomous system is not subject to any external influences that depend on the time, so the velocity function is independent of time and the equation of motion is

$$\dot{x} = v(x). \tag{1.2}$$

The conditions which determine the motion of an autonomous system are independent of time, so we will sometimes refer to it as a system with *time-independent conditions*.

If x_0 represents the state at time t_0, then (1.2) gives t as a function of x:

$$t - t_0 = \int_{x_0}^{x} \frac{dx}{v(x)},$$ (1.3)

if the integral exists. The inverse function gives x as a function of t. Notice that x depends on t and t_0 only through the time interval $t - t_0$. This property is restricted to autonomous systems.

Without solving equation (1.2) or integrating (1.3), we can obtain the qualitative behaviour of the system graphically, as illustrated in the top half of figure 1.1 in which we represent the velocity function $v(x)$ (we have chosen $v(x) = -x + x^3$ for the sake of example) by a set of arrows as follows:

Fig. 1.1. Graphs of $v(x) = -x + x^3$ and $U(x) = \frac{1}{2}x^2 - \frac{1}{4}x^4$ with arrows representing the phase flow.

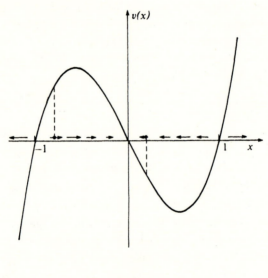

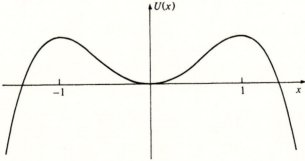

take a suitable set of values, x_s, of x and, for each x_s, draw an arrow of length proportional to $v(x_s)$ on or near the x-axis, with its *centre* at x_s and pointing in the direction of increasing or decreasing x, depending on the sign of $v(x_s)$.

We can think of a fluid flowing in the phase space with velocity $v(x)$ at all times. The arrows represent the velocity of the fluid, named the *phase flow* and $v(x)$ is its *velocity field*. In this example the fluid is clearly compressible. The changing state of the system is like a particle carried along by the fluid.

The velocity field $v(x)$ of an autonomous first-order system can be expressed as the negative gradient of a potential $U(x)$. Thus

$$v(x) = -\frac{dU}{dx},$$ (1.4a)

where, for some constant U_0,

$$U(x) = U_0 - \int_0^x dx' \, v(x').$$ (1.4b)

The states of the system flow 'downhill' away from the maxima of the potential $U(x)$, like water flowing down hills and into valleys, as illustrated in the lower half of figure 1.1.

At each zero x_k of the velocity field $v(x)$,

$$v(x_k) = 0,$$ (1.5)

so that a system initially at x_k remains there for all time. The points x_k represent states of *equilibrium*: they are named *fixed points*. At all other points the state of the system changes. A system in an open interval between two fixed points cannot pass either of them. Such open intervals, together with those that extend from a fixed point to infinity, are invariant, as are the fixed points. Such fixed points and intervals represent *invariant sets of states* which are defined by the property that if any system is in such a set at some time, then it remains in that set for all times. We usually consider only those elementary invariant sets which cannot be decomposed into smaller invariant sets.

The whole of the phase space of a system is made up of invariant sets, and they provide valuable information about behaviour over arbitrarily long periods of time. The system illustrated in figure 1.1 has three fixed points at $x = 0, \pm 1$, and four elementary one-dimensional invariant sets, which are the intervals $(-\infty, -1)$, $(-1, 0)$, $(0, 1)$, $(1, \infty)$, bounded on one or two sides by the fixed points.

When the velocity function $v(x)$ has only simple zeros, the fixed points are of two types. There are the *stable* points x_k around which $v(x)$ is a decreasing function of x, so that neighbouring states approach x_k, and there are *unstable* points x_k around which $v(x)$ is an increasing function of x so that neighbouring states leave x_k, as shown in figure 1.2.

Fig. 1.2. Typical velocity fields $v(x)$, potential $U(x)$ and flows in the neighbourhood of stable and unstable fixed points.

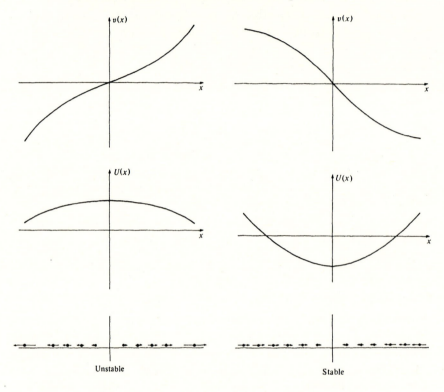

Unstable   Stable

Example 1.1

The velocity field is $v(x) = 0$. Every point is a fixed point. The system is always and everywhere at rest,

$$x = x_0. \tag{1.6}$$

Every phase point and every set of phase points is an invariant set.

Example 1.2

The velocity field is $v(x) = a, a \neq 0$. If $x = x_0$ at $t = 0$,

$$x = x_0 + at \tag{1.7}$$

with phase diagram

$a > 0$ (1.8)

This is the case of uniform flow, and the entire phase space $(-\infty, \infty)$ is the only invariant set.

Example 1.3

The velocity field is $v(x) = bx$, $b \neq 0$. Then

$$x = x_0 e^{bt} \tag{1.9}$$

with phase diagrams

$b < 0$

$$\tag{1.10}$$

$b > 0$

There is one fixed point at $x = 0$, which is stable for $b < 0$ and unstable for $b > 0$.

This example is particularly important as it describes the motion in a sufficiently small neighbourhood of most fixed points. An exception is given by the next example. The first three examples are of *linear* systems, whose motion satisfies a linear differential equation.

Example 1.4

The velocity field is

$$v(x) = cx^2 \tag{1.11a}$$

with phase diagram

$c > 0$ (1.11b)

There is one fixed point at $x = 0$, but it is both stable and unstable. Neighbouring states approach it from the left, but leave it from the right. This is because $x = 0$ is a double zero of the velocity function.

The velocity field (1.11) has a property known as *structural instability*. If we take the previous two examples and add to the velocity field a small arbitrary perturbing field $\epsilon w(x)$, where $w(x)$ is bounded and differentiable, then, for sufficiently small ϵ, the structure of the invariant sets remains unchanged. But if we do the same for the field (1.11), for example with $w(x) = 1$, the number of zeros of $v(x)$ is increased to 2 or decreased to 0, however small ϵ may be. A system which retains its structure under such a small perturbation is *structurally stable*. Otherwise it is structurally unstable, like example 1.4.

Care must be taken to distinguish the stability of states, or phase points, from the structural stability of systems, or differential equations. For simplicity we usually consider structurally stable systems, because they are typical.

The system with velocity field (1.11) also illustrates another important feature, *terminating motion*. Suppose at time $t = 0$, $x = x_0 > 0$. Then the solution of the differential equation of motion gives

$$x = (x_0^{-1} - ct)^{-1}, \quad t < (cx_0)^{-1}. \tag{1.12}$$

The motion terminates abruptly at the critical time $t = (cx_0)^{-1}$ when x tends to infinity, and is undefined beyond that time. Not all velocity fields define the motion of the system for all time. In practice the validity of the equation of motion itself breaks down before the critical time is reached. For $x_0 < 0$ there is no positive critical time, but an attempt to determine the *past* motion of the system leads to a similar difficulty.

In general, we refer to the motion as *terminating* if it is represented by a solution of the equation of motion which is undefined at any finite point of the real time axis. Sometimes, but not always, this happens when x becomes infinite. Terminating motion is typical of systems in the exterior invariant intervals of certain polynomial velocity fields. It occurs whenever $v(x)$ is a polynomial of degree two or greater.

1.2   Rotation

The phase space does not always occupy the whole real line. For some types of motion, typified by rotation about an axis, the phase space can be considered as a circle. In that case it is convenient to choose a coordinate θ in the range $[0, 2\pi]$ and to identify the ends of the range. The equation of motion for an autonomous system is

$$\dot{\theta} = v(\theta), \tag{1.13}$$

where $v(\theta)$ must be a periodic function of θ with period 2π, $v(\theta) = v(\theta + 2\pi)$. For these systems it is possible to have bounded motion with no fixed points, for example

$$\dot{\theta} = \omega = \text{constant}, \tag{1.14a}$$

which represents uniform rotation with period $T = 2\pi/\omega$.

(1.14*b*)

Any circular motion with no fixed point is named a *rotation* and has a period T given by

$$T = \int_{\theta=0}^{\theta=2\pi} \mathrm{d}t = \int_{0}^{2\pi} \frac{\mathrm{d}\theta}{v(\theta)}.$$

(1.15)

1.3   Natural boundaries

If the variable x represents the distance of a particle from a point in space, or the population of a large number of objects such as living cells or radioactive atoms, then negative values of x have no meaning. The phase space has a *natural boundary* at $x = 0$. Usually it helps to include the boundary point in the phase space. Normally, for first-order systems, if the motion does not terminate, the velocity at a natural boundary is zero.

For first-order autonomous systems examples of velocity fields with natural boundaries are given by taking any velocity field with fixed points and restricting the phase space to an invariant set bounded by, and including, a fixed point or points. For example, (1.10), excluding the negative real axis, represents exponential decay of a population, with a natural boundary at $x = 0$.

Example 1.5 A system that terminates at its natural boundary.
The equation of motion is

$$\dot{x} = v(x) = -\sqrt{x} \quad (x \geqslant 0),$$

(1.16)

where the square root is taken as positive. The system has a natural boundary at $x = 0$, and x always decreases. The general solution of the equation of motion is given by $\mathrm{d}x/\sqrt{x} = -\mathrm{d}t$; it is

$$2\sqrt{x} = C - t,$$

where C is a constant, so

$$x = \tfrac{1}{4}(C-t)^2, \quad (t \leqslant C).$$

(1.17)

The system reaches the natural boundary at $t = C$. Later than this, equation (1.17) does not satisfy (1.16) but is a solution of

$$\dot{x} = \sqrt{x}, \qquad (1.18)$$

with the opposite sign. No motion satisfying (1.16) is possible beyond $t = C$. However equation (1.17) defines a possible motion in which the system satisfies the equation (1.16) for $t \leqslant C$ and (1.18) for $t \geqslant C$. Hamiltonian systems with conserved quantities sometimes move in this way.

More general systems and theorems of this type appear in exercises 1.8 and 1.9.

1.4 Examples from biology

The dynamics of biological populations is a branch of ecology. The populations of insects, birds, fish and mammals are increased by birth and decreased by death. These processes depend on many factors that are less well understood than the factors that influence the motion of levers, pulleys, projectiles and planets. Nevertheless, simplified models can help us to understand the way that these populations change with time.

When the population of a single species in some specified region is sufficiently large it may be represented by a continuous variable x. If the birth and death rates per individual of the populations, $B(x)$ and $D(x)$, depend on the population x, but not on space, time, or any other factors, then the population is a first-order autonomous system with equation of motion $\dot{x} = v(x)$ and velocity function

$$v(x) = [B(x) - D(x)]\, x \qquad (B(x) \geqslant 0, \; D(x) \geqslant 0). \qquad (1.19)$$

For populations, $x \geqslant 0$, so that $x = 0$ is always a natural boundary.

Example 1.6 Exponential growth and decline

The simplest assumption is that the birth rate B and death rate D are constants independent of the population. The equations are those of example 1.3 with $b = B - D$. If we exclude the special case when $B = D$, the natural boundary at $x = 0$ is the only fixed point. If D exceeds B then the population decays exponentially to zero. If B exceeds D, the population increases exponentially without bound. The phase diagrams are given by the right halves of (1.10).

Example 1.7 The logistic equation

In practice the population in a confined region cannot increase without bound forever, as there are limiting factors, such as competition for food and living space. The simplest assumption is to suppose that these factors leave the birth rate unchanged, but produce a death rate per individual proportional to the population, so we can write

$$B(x) = b, \quad D(x) = cx \qquad (x \geqslant 0, b > 0, c > 0) \qquad (1.20)$$

and the equation of change has the form

$$\dot{x} = bx - cx^2 . \tag{1.21}$$

This is known as the *logistic equation* and many actual populations closely follow it. It is not difficult to solve, or to analyse, so this is given as exercise 1.13. It gives a stable population.

However, the situation is rarely so simple. Frequently one species preys on another, so that their population equations are coupled together, leading to systems of second order as in chapter 3.

Sometimes an individual species has a definite reproductive season, so that the change in population is not represented by a differential equation, but by a difference equation or map. This is considered briefly in chapter 11. In both cases completely new and remarkable phenomena appear in the time dependence of the populations.

Exercises for chapter 1

(1) Three first-order systems have the following velocity functions. Which of them is autonomous?

    (a) $v(x, t) = e^x$;

    (b) $v(x, t) = t$;

    (c) $v(x, t) = \begin{cases} 0 & (t < 0) \\ x^2 & (t \geqslant 0) . \end{cases}$

(2) Draw the phase diagrams and find the fixed points and invariant sets of systems with the following velocity functions:

    (a) $v(x) = (a - x)(x - b)$ $(-\infty < x < \infty, b > a > 0)$;

    (b) $v(x) = (a - x)(b - x)$ $(-\infty < x < \infty, b > a > 0)$.

Without solving the equations of motion, discuss the qualitative behaviour after $t = 0$ if $a < x(0) < b$ in each case.

(3) The angle of a blade of a food mixer is denoted by ψ and its motion is determined by the differential equation

$$\dot{\psi} = a + b \sin \psi \quad (0 \leqslant \psi \leqslant 2\pi, a > 0, b > 0).$$

Find the fixed points and invariant sets of the motion when $a < b, a = b$ and $a > b$. For what values of a and b is the motion a rotation? Find the period.

(4) Is the system of example 1.1 structurally stable? Give reasons for your answer.

(5) For what positive values of a and b is the system with velocity function

$$v(x) = (a - x)(x - b) \quad (-\infty < x < \infty, b > 0, a > 0)$$

structurally stable?

(6) For what values of a and b is the food mixer blade of exercise 1.3 structurally stable?

(7) Discuss the nature of the motion of the systems with velocity functions

    (a) $v(x) = x \sin x \quad (-\infty < x < \infty)$,

    (b) $v(x) = x \cos x \quad (-\infty < x < \infty)$,

    and determine which of them, if any, is structurally stable.

(8) For what positive values of α does the motion with velocity function

$$v(x) = x^\alpha \quad (x > 0, \alpha > 0)$$

terminate?

Exercises for Hamiltonian mechanics (all square roots are non-negative)

(9) State the important features of the motion of the systems with velocity functions

    (a) $v(x) = a\sqrt{x - b} \quad (a > 0, x \geqslant b)$,

    (b) $v(x) = \sqrt{1 - x^2} \quad (|x| \leqslant 1)$.

(10) (a)   Suppose that $A < B$ and that

$$0 < v_1(x) \leqslant v_2(x) \quad (A \leqslant x \leqslant B)$$

$v_1(B), v_2(B)$ finite.

Show that, if T_i is the time that system S_i with velocity function v_i takes to go from state $x = A$ to state $x = B$, then

$$T_2 \leqslant T_1.$$

    (b)   Hence show that, if $f(x)$ and $f'(x)$ exist in an interval around $x = B$ with values satisfying

$$f(B) = 0, f'(B) < 0,$$

then the motion defined by

$$v(x) = \sqrt{f(x)}$$

terminates at $x = B$.

    (c)   Describe the motion with velocity function

$$w(x) = \sqrt{g(x)},$$

where $g(x)$ is a polynomial with N distinct real zeros.

Exercises on chemical reactions

(11) Nitric oxide (NO) and oxygen (O_2) react to form NO_2 as follows

$$2NO + O_2 \rightarrow 2NO_2.$$

If $C(t)$ denotes the concentration of NO_2, it is found to satisfy the differential equation

$$dC/dt = k(\alpha - C)^2 (2\beta - C) \quad (C \geqslant 0, C(0) = 0) \,;$$

where k is a positive constant for the reaction and α and β are the initial concentrations of NO and O_2 respectively, both greater than zero. Discuss the qualitative behaviour of the NO_2 concentration when $\alpha < 2\beta$ and $\alpha > 2\beta$.

(12) Sulphur dioxide reacts with oxygen to form SO_3

$$2SO_2 + O_2 \rightarrow 2SO_3.$$

The concentration $C(t)$ of SO_3 satisfies the equation

$$dC/dt = k (\alpha - C) C^{-\frac{1}{3}} \quad (C \geqslant 0, C(0) = 0),$$

where k is a positive constant for the reaction and $\alpha > 0$ is the initial concentration of SO_2. Discuss the change in concentration $C(t)$ for small times and large times.

Exercises on biology

(13) Draw the phase diagram for a population satisfying the logistic equation (1.21) and describe what happens when it starts at any positive value.

(14) Various modifications of the logistic equation have been suggested. Suppose a population obeys the differential equation

$$\dot{x} = ax(1 - b(e^x - 1)) \quad (x > 0, a > 0, b > 0).$$

Describe qualitatively the change of population with time.

Additional exercises

(15) A spherical drop of fluid loses mass by evaporation at a rate proportional to its surface area. Determine the differential equation for the radius $R(t)$ as a function of time, and describe the change of radius qualitatively.

(16) With air resistance proportional to the square of the velocity, the velocity $V(t)$ of a falling body is given by the equation

$$dV/dt = g - kV^2 \quad (g > 0, k > 0).$$

Describe the change of $V(t)$ with time, for arbitrary initial non-negative values of $V(0)$. What is the limiting behaviour as $t \rightarrow \infty$?

2   LINEAR TRANSFORMATIONS OF THE PLANE

2.1   Introduction

When we come to the theory of second-order systems we will need many properties of linear transformations of a plane and of their matrix representations. These are summarized for convenience in this chapter.

Consider a real plane with rectangular Cartesian coordinate system Oxy. A linear transformation or linear map of the plane onto itself that leaves the origin unchanged may be represented by a real 2×2 matrix

$$A = \begin{pmatrix} a & b \\ c & d \end{pmatrix}. \tag{2.1}$$

A point with position vector $\mathbf{r} = (x, y)$ is mapped onto the point $\mathbf{R} = (X, Y)$, where

$$\begin{pmatrix} X \\ Y \end{pmatrix} = \begin{pmatrix} a & b \\ c & d \end{pmatrix} \begin{pmatrix} x \\ y \end{pmatrix}. \tag{2.2a}$$

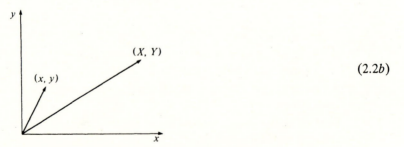

$$(2.2b)$$

The transformation of any geometrical figure is obtained from the transformation of the points which constitute it. Because the transformation is linear, straight lines are transformed into straight lines. For example, a square may be transformed as shown in (2.3).

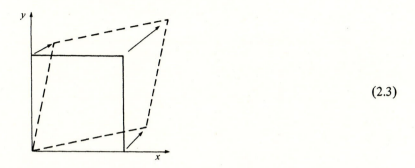

(2.3)

The area of any region may or may not be changed by a linear transformation.

We require the transformation to be invertible, so the determinant,

$$\delta = \text{Det } A = ad - bc,$$ (2.4)

must not be zero. The ratio of areas is δ, so if the transformation preserves area, the determinant is unity,

$$\delta = 1 \iff \text{area-preservation.}$$ (2.5)

Area-preserving transformations are particularly important in Hamiltonian dynamics.

Linear transformations may be classified into distinct types, depending on the eigenvalues of A, λ_1, λ_2, that are the solutions of the characteristic equation

$$\text{Det } (A - \lambda I) = 0,$$ (2.6)

where I is the unit matrix. This equation may be rewritten as

$$\lambda^2 - \tau\lambda + \delta = 0,$$ (2.7)

where τ is the trace of A

$$\tau = \text{Tr } A = a + d.$$ (2.8)

For area-preserving transformations the classification depends only on the trace τ.

The eigenvalues are related to the determinant and trace by

$$\lambda_1 + \lambda_2 = \tau, \quad \lambda_1 \lambda_2 = \delta.$$ (2.9)

2.2   Area-preserving transformations

The solutions of the characteristic equation (2.7) for the area-preserving transformations with $\delta = 1$ are given by

$$\lambda = \tfrac{1}{2}\tau \pm [(\tfrac{1}{2}\tau)^2 - 1]^{\tfrac{1}{2}}$$ (2.10)

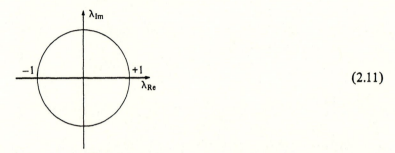

(2.11)

They always lie on the real axis or the unit circle in the complex plane. Where they lie depends on the value of the trace τ. The value of τ or the positions of the eigenvalues can be used to classify the transformations into three distinct types.

Type 1 $|\tau| > 2$ and the eigenvalues are real and distinct

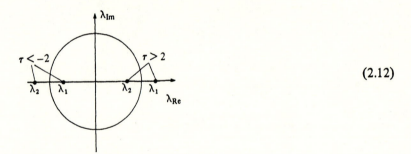

(2.12)

To be definite, we always choose $\lambda_1 > \lambda_2$.

Type 2 $|\tau| < 2$ and the eigenvalues are complex conjugates

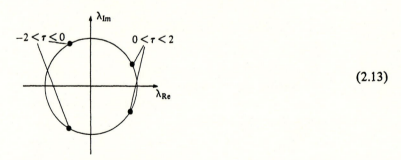

(2.13)

In this case we often write the eigenvalues as

$$\lambda_1 = \alpha + i\omega = e^{i\theta} \qquad \left\{ \begin{array}{l} \alpha, \omega, \theta \text{ real} \\ 0 < \theta < \pi \end{array} \right\} \hspace{1cm} \left. \right\} \hspace{1cm} (2.14)$$

$$\lambda_2 = \alpha - i\omega = e^{-i\theta}$$

$$\text{Sgn } \omega = \text{Sgn } c.$$

To be definite, and for later convenience, we choose the sign of ω to be the same as the sign of the element c of the matrix A.

Type 3 $|\tau| = 2$ and the eigenvalues are real and equal

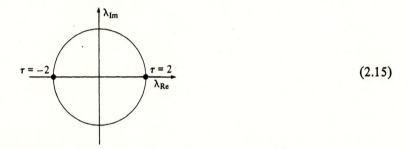

$$(2.15)$$

The eigenvalues are

$$\lambda_1 = \lambda_2 = 1 \text{ or } \lambda_1 = \lambda_2 = -1. \hspace{2cm} (2.16)$$

Usually the special case when A equals the unit matrix is excluded from type 3.

The matrix A representing the transformation can be simplified by making a linear change of coordinates, represented by a non-singular matrix M. According to the standard theory this changes the matrix A to the matrix B given by

$$B = MAM^{-1}. \hspace{2cm} (2.17)$$

Because the determinant of a product is the product of the determinants, the eigenvalues of B are the same as those of A, so the type of transformation represented by B is the same as the type for A. For each type we can choose a matrix M, depending on the elements of A, so that B is of a simple standard form. The standard form is different for each type, and represents a different standard type of linear transformation.

We illustrate these standard transformations by their effect on the unit square with corners at

$$(0, 0), (0, 1), (1, 1), (1, 0).$$

For the matrix A with elements a, b, c, d given by equation (2.1) and eigenvalues λ_1, λ_2, we write down the transformation matrix, M, the standard form, B, and the effect on the unit square.

For type 1, real distinct eigenvalues, $\lambda_1 > \lambda_2$, they are

$$M = \begin{pmatrix} c & \lambda_1 - a \\ c & \lambda_2 - a \end{pmatrix}, \quad B = \begin{pmatrix} \lambda_1 & 0 \\ 0 & \lambda_2 \end{pmatrix}, \tag{2.18}$$

with the following effect on the unit square

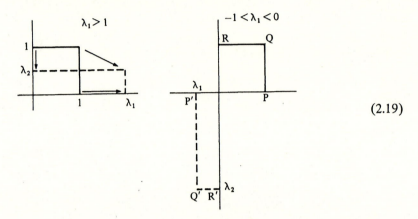

$$(2.19)$$

For positive λ_i the transformation by B is a *standard squeeze* and the transformation by A is a *squeeze*. For negative λ_i there is also a reflexion in the origin.

For type 2, complex eigenvalues, $\lambda = \alpha \pm i\omega = e^{\pm i\theta}$, they are

$$M = \begin{pmatrix} c & \alpha - a \\ 0 & \omega \end{pmatrix}, \quad B = \begin{pmatrix} \cos\theta & -\sin\theta \\ \sin\theta & \cos\theta \end{pmatrix}, \tag{2.20}$$

and the effect on the unit square is shown below

$$(2.21)$$

The transformation by B is a *rotation* through an angle θ in the range $-\pi < \theta < \pi$ and the transformation by A is a *generalized rotation*.

For type 3, real and equal eigenvalues, they are

$$M = \begin{pmatrix} a - d & 2b \\ 2c & 0 \end{pmatrix}, \quad B = \begin{pmatrix} \lambda & 0 \\ c & \lambda \end{pmatrix} \quad (\lambda = \pm 1), \tag{2.22}$$

with the effect shown below.

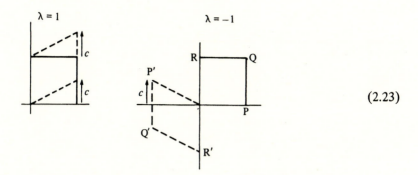

(2.23)

Either $\lambda = 1$ and the transformation by B is a *standard shear* in the y-direction or $\lambda = -1$ and the transformation is a reflexion in the origin followed by a shear. The special case when b or $c = 0$ is considered in exercise 2.5. The transformation by A is a *shear*.

Example 2.1
Of what type is the transformation represented by the matrix

$$A = \begin{pmatrix} 2 & 0 \\ \frac{3}{2} & \frac{1}{2} \end{pmatrix}$$

(2.24)

Obtain the standard form B and describe the type of transformation that it represents. Sketch the points $B^n \binom{1}{1}$ and $B^n \binom{-1}{1}$ for integer n with $-2 \leqslant n \leqslant 2$.

Since $\delta = 1$ the transformation is area-preserving and since $\tau = \frac{5}{2} > 2$ it is of type 1. The matrix B is formed from the eigenvalues

$$B = \begin{pmatrix} 2 & 0 \\ 0 & \frac{1}{2} \end{pmatrix}$$

(2.25)

and represents a squeeze. The matrix B^n is given by

$$B^n = \begin{pmatrix} 2^n & 0 \\ 0 & 2^{-n} \end{pmatrix}$$

(2.26)

which is a squeeze by a factor 2^n. The points obtained by operating with B^n on $\binom{1}{1}$ and $\binom{-1}{1}$ are represented in the sketch below.

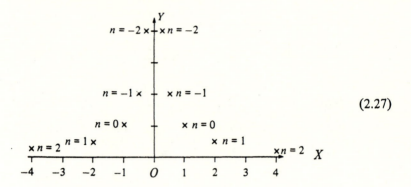

$$(2.27)$$

2.3   Transformations with dilation

This section is not needed for Hamiltonian dynamics.

When A is not area-preserving, the eigenvalue solutions λ of the characteristic equation (2.7) are given in terms of the trace τ and determinant δ by

$$\lambda = \tfrac{1}{2}\tau \pm [(\tfrac{1}{2}\tau)^2 - \delta]^{\frac{1}{2}} \tag{2.28}$$

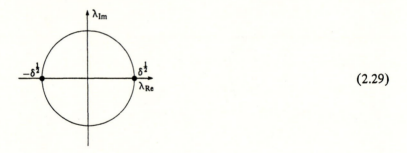

$$(2.29)$$

If $\delta > 0$ they lie on the real axis, or on the circle

$$|\lambda|^2 = \delta \quad (\delta > 0) \tag{2.30}$$

in the complex plane. If $\delta < 0$ they always lie on the real axis.

The different types are distinguished by the conditions set out below.

Type 1: $\tau^2 > 4\delta$: eigenvalues real and distinct. $\tag{2.31}$

Type 2: $\tau^2 < 4\delta$: eigenvalues complex conjugate. $\tag{2.32}$

Type 3: $\tau^2 = 4\delta$: eigenvalues real and equal. $\tag{2.33}$

For $\delta > 0$ the diagrams and equations (2.12)–(2.16) can be used, provided that we remember that the circles in the complex plane have radius $\delta^{\frac{1}{2}}$ as in (2.29), that τ must be replaced by $\tau/\delta^{\frac{1}{2}}$ and that equations (2.14) are replaced by

$$\lambda_1 = \alpha + i\omega = \delta^{\frac{1}{2}} e^{i\theta} \quad \left\{ \begin{array}{l} \alpha, \omega, \theta \text{ real} \\ 0 < \theta < \pi \end{array} \right\}.$$
$$\lambda_2 = \alpha - i\omega = \delta^{\frac{1}{2}} e^{-i\theta}$$
(2.34)

For negative determinant δ, A must be of type 1, and the eigenvalues are real with $\lambda_1 > 0$, $\lambda_2 < 0$.

For all non-singular real A, the matrices M are exactly the same as they are for the area-preserving transformations, as given in equations (2.18), (2.20), (2.22), but the standard forms B are not always the same. However they can easily be found. If A has positive determinant δ, then we can write

$$A = \delta^{\frac{1}{2}} A_{\text{AP}} \quad (\delta > 0),$$
(2.35)

where A_{AP} is area-preserving. Therefore

$$A = \delta^{\frac{1}{2}} M B_{\text{AP}} M^{-1} = M(\delta^{\frac{1}{2}} I B_{\text{AP}}) M^{-1},$$
(2.36)

where B_{AP} is in one of the three standard forms for an area-preserving transformation and

$$\delta^{\frac{1}{2}} I = \begin{pmatrix} \delta^{\frac{1}{2}} & 0 \\ 0 & \delta^{\frac{1}{2}} \end{pmatrix}$$
(2.37)

This matrix represents a *dilation*. It multiplies all linear dimensions by a factor $\delta^{\frac{1}{2}}$ and all areas by δ. The effect on the unit square is illustrated in (2.38).

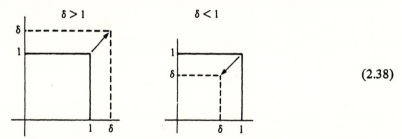

(2.38)

A dilation by a factor of less than unity is commonly known as a contraction, but it suits our purposes to use the same word for any positive value of δ. All matrices with negative δ are of type 1 and are considered in the next paragraph.

The matrices B and the standard transformations for the three types are as follows.

Type 1

$$B = \begin{pmatrix} \lambda_1 & 0 \\ 0 & \lambda_2 \end{pmatrix} \quad (\lambda_1, \lambda_2 \text{ real and different})$$
(2.39)

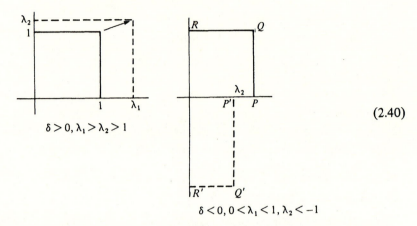

$$\delta > 0, \lambda_1 > \lambda_2 > 1$$

$$\delta < 0, 0 < \lambda_1 < 1, \lambda_2 < -1$$

(2.40)

For $\lambda_1 > \lambda_2 > 1$ the standard transformation is a squeeze and dilation, or an *anisotropic dilation*. All x components are multiplied by λ_1, y components by λ_2. For one negative eigenvalue ($\delta < 0$) the standard transformation is a squeeze, dilation and reflexion in an axis. For two negative eigenvalues ($\delta > 0$) it is a squeeze, dilation and reflexion in the origin.

Type 2

$$B = \begin{pmatrix} \alpha & -\omega \\ \omega & \alpha \end{pmatrix} = \delta^{\frac{1}{2}} \begin{pmatrix} \cos\theta & -\sin\theta \\ \sin\theta & \cos\theta \end{pmatrix}.$$ (2.41)

The standard transformation is a rotation and dilation.

Type 3

$$B = \begin{pmatrix} \lambda & 0 \\ c & \lambda \end{pmatrix} \quad (\lambda = \pm\delta^{\frac{1}{2}}).$$ (2.42)

The standard transformation is a shear and dilation for $\lambda = \delta^{\frac{1}{2}}$, with a reflexion in the origin also, if $\lambda = -\delta^{\frac{1}{2}}$.

Example 2.2
Of what type is the transformation by the matrix

$$A = \begin{pmatrix} 1 & 1 \\ -1 & 1 \end{pmatrix}$$ (2.43)

and what is the standard form B ? What is the name of the transformation produced by B ? Sketch the points $B^n\begin{pmatrix}1\\0\end{pmatrix}$ for integer n, $-4 \leqslant n \leqslant 4$.

The determinant is $\delta = 2$, so the transformation by A is not area-preserving and involves a dilation. The trace is $\tau = 2$ so, by (2.32), it is of type 2 and, by

(2.28), the eigenvalues are

$$\lambda = 1 \pm (-1)^{\frac{1}{2}} = 1 \pm i .$$ (2.44)

Since the lower left element of A is negative, so is ω, and

$$\alpha + i\omega = 1 - i$$

$$= \delta^{\frac{1}{2}} (\cos \theta + i \sin \theta),$$ (2.45)

so $\quad \cos \theta = \dfrac{1}{\sqrt{2}}, \;\; \sin \theta = -\dfrac{1}{\sqrt{2}}, \;\; \theta = -\tfrac{1}{4}\pi .$

The matrix

$$B = \begin{pmatrix} 1 & 1 \\ -1 & 1 \end{pmatrix} = \sqrt{2} \begin{pmatrix} \cos(-\tfrac{1}{4}\pi) & -\sin(-\tfrac{1}{4}\pi) \\ \sin(-\tfrac{1}{4}\pi) & \cos(-\tfrac{1}{4}\pi) \end{pmatrix}$$ (2.46)

is the same as A, and represents a rotation through an angle $-\tfrac{1}{4}\pi$ (a clockwise rotation of $45°$) and a dilation of $\sqrt{2}$. Without further calculation B^n represents a rotation through $-\tfrac{1}{4}n\pi$ and a dilation of $2^{n/2}$, so the nine requested points are

(2.47)

Exercises for chapter 2

(1) Obtain the type and standard form, B, of the following matrices, A. State what kind of transformation B represents and whether it is area-preserving.

(a) $\begin{pmatrix} -1 & 4 \\ -2 & 3 \end{pmatrix}$;   (b) $\begin{pmatrix} 1 & 1 \\ 1 & 2 \end{pmatrix}$;

(c) $\begin{pmatrix} -1/\sqrt{5} & -4/\sqrt{5} \\ 2/\sqrt{5} & 3/\sqrt{5} \end{pmatrix}$;   (d) $\begin{pmatrix} 1 & -1 \\ 1 & 3 \end{pmatrix}$:

(2) Show that if $B = MAM^{-1}$ then Det B = Det A and the eigenvalues of A and B are the same.

(3) Of what type is the transformation represented by the matrix

$$A = \frac{1}{2} \begin{pmatrix} \sqrt{3} & -1 \\ 1 & \sqrt{3} \end{pmatrix}$$

and what is the standard form B? Does B represent a squeeze, a rotation or a shear? Sketch the points $B^n \begin{pmatrix} 1 \\ 0 \end{pmatrix}$, for integer n, in the (X, Y) plane.

(4) Show that if a real matrix M with positive determinant can be found that converts A to standard form B by equation (2.17), then a real matrix M', representing an area-preserving transformation, can be found with the same properties. Obtain such an M' explicitly in terms of the elements and eigenvalues of A for types 1 and 2.

(5) Show that if

$$A = \begin{pmatrix} a & b \\ c & d \end{pmatrix}$$

is of type 3 and $b = 0$ it must already be in standard form so M can be the unit matrix. Show that if $c = 0, b \neq 0$, then the matrix

$$M = \begin{pmatrix} 0 & -1 \\ 1 & 0 \end{pmatrix}$$

converts A to standard form.

(6) Obtain the matrices that represent:

    (*a*) a rotation through an angle $120°$;

    (*b*) a standard squeeze that transforms $\begin{pmatrix} 1 \\ 0 \end{pmatrix}$ to $\begin{pmatrix} 3 \\ 0 \end{pmatrix}$.

(7) Show that, if A represents a general rotation, then $b \neq 0$ and $c \neq 0$.

3   SECOND-ORDER AUTONOMOUS SYSTEMS

3.1   Systems of order n

Most systems are more complicated than the first-order systems described in chapter 1. With the exception of very special limiting cases, all Newtonian mechanical systems are of higher order than the first, as is any system with more than one component, or more than one direction of motion.

A system of *order n* is defined by two properties:

(N1) The state of the system is represented by n real variables, $x_1, x_2, \ldots x_n$, or one real vector variable r of dimension n, which may be considered as coordinates of an abstract n-dimensional space named the *phase space*.

(N2) The motion of the system is represented by a vector function $r(t)$ of time satisfying a first-order vector differential equation of motion

$$\frac{dr}{dt} = \dot{r} = v(r, t), \tag{3.1}$$

where v is a given, sufficiently well-behaved vector *velocity function* of r and t, whose value for a particular r and t is the *phase velocity*.

The differential equation (3.1) is the *equation of motion* or *equation of change* of the system. The conditions on v required for the existence and uniqueness of the solution (3.1) are given in appendix 1.

When the velocity function is independent of t the system is autonomous. The solutions then depend upon t and the initial time t_0 only through the difference $t - t_0$. For systems of higher order than the first there is usually no solution like (1.3) so qualitative methods are more important than for first-order systems.

During a particular motion of the system, described by a function $r(t)$, the states trace out a continuous curve in the phase space named a *phase curve*. For most of our systems the motion at all times is determined by the state at a particular time, usually chosen to be $t = 0$. In that case the differential equations (3.1) possess a unique solution $r(t)$, for given value of $r(0)$, that is, given initial conditions. The exceptions are systems with terminating motion. The set of all possible motions is named the *phase flow*.

For systems of second order, the phase space could be a plane, or part of a plane, or a cylinder, or a more complicated surface. We suppose at first that it is a plane, and that the vector has Cartesian components x, y so that

$$\mathbf{r}(t) = (x(t), y(t)). \tag{3.2}$$

In terms of components, the vector differential equation becomes

$$\frac{\mathrm{d}x}{\mathrm{d}t} = \dot{x} = v_x(x, y, t),$$
$$\frac{\mathrm{d}y}{\mathrm{d}t} = \dot{y} = v_y(x, y, t) \tag{3.3}$$

which are two coupled first-order differential equations for $x(t)$ and $y(t)$. If the system is autonomous, t does not appear as an argument of v_x or v_y and then a differential equation for the phase curve is obtained by eliminating time from equation (3.3) to give

$$\frac{\mathrm{d}y}{\mathrm{d}x} = \frac{v_y(x, y)}{v_x(x, y)} . \tag{3.4}$$

As an example, a particle constrained to move in one dimension under action of a force F which depends on the position x of the particle, but not on the time, nor on the velocity, satisfies Newton's equation of motion:

$$m \frac{\mathrm{d}^2 x}{\mathrm{d}t^2} = m\ddot{x} = F(x). \tag{3.5}$$

If we put $y = \dot{x}$, $\mathbf{r} = (x, y)$ then we can rewrite the second-order differential equation (3.5) as two coupled first-order equations,

$$\dot{x} = y, \quad \dot{y} = F(x)/m , \tag{3.6}$$

or one vector equation,

$$\dot{\mathbf{r}} = (y, F(x)/m) , \tag{3.7}$$

which have the standard forms (3.3) and (3.1) with phase velocity

$$v(x, y) = (y, F(x)/m) . \tag{3.8}$$

The system is therefore of second order. Note that, whereas the first element of the phase velocity vector is a physical velocity, the second element is a physical acceleration.

In general, any finite set of coupled differential equations that defines the motion of a system may be expressed in the standard form (3.1) by defining higher derivatives than the first as new dependent variables, as for our special example. The minimum number of coupled first-order equations obtained by this process is the *order* of the system. An unconstrained Newtonian system of N degrees of freedom is a system of order $2N$.

3.2 Phase flows of second-order autonomous systems

We now restrict ourselves to second-order autonomous systems and represent the possible motions of a system by a diagram on the phase plane known as a *phase diagram*, or sometimes as a *phase portrait*, of the phase flow. In this way we are able to obtain the qualitative behaviour of the system without explicit solution. The procedure is similar to that for first-order systems, to which the reader is referred.

Represent the velocity function $v(\mathbf{r})$ by a set of arrows whose magnitude and direction are proportional to the local velocity, as follows. Take a suitable mesh of points on the phase plane, and through each point P draw a short arrow with its centre at P to denote the velocity vector. The length of the arrow is proportional to the magnitude of the velocity and its direction is the same.

Example 3.1

Consider the particular case of a particle falling freely in a vertical direction in the Earth's gravitational field, assumed to be uniform. The phase velocity vector is

$$v(x, y) = (y, -g) \tag{3.9}$$

and the phase diagram is

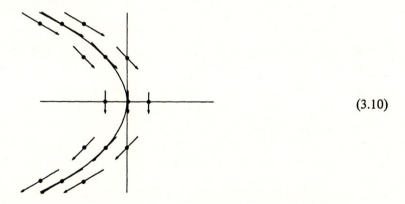

$$(3.10)$$

The arrows represent the phase flow, and overall the diagram looks like a representation of the flow of a fluid. An individual orbit is represented by the motion of a particle carried along by the fluid. A phase curve can be drawn by following the arrows, interpolated where necessary. The phase curve should touch its tangent arrows at their centres.

The example we have chosen is one of those exceptional systems for which an analytic solution is possible. The equations of motion (3.6) may be written

$$\frac{dx}{dt} = y, \quad \frac{dy}{dt} = -g, \tag{3.11}$$

with solutions

$$x(t) = x_0 + y_0\, t - \tfrac{1}{2} g t^2, \quad y(t) = y_0 - g t, \tag{3.12}$$

where (x_0, y_0) is the phase point at $t = 0$. The vector function $(x(t), y(t))$ defines the motion.

From equation (3.4) the phase curves satisfy the equation

$$\frac{dy}{dx} = -g/y, \tag{3.13}$$

with solutions

$$x = x_1 - y^2/2g, \tag{3.14}$$

where x_1 is the height attained when $y = 0$. Diagram (3.10) shows a phase curve for this motion. It must be emphasized that analytic solutions are not usually obtainable, so the analytic method can rarely be used.

However, *local* analytic solutions, which are approximately valid over part of phase space, are generally applicable and very useful, particularly in the neighbourhood of the fixed points of the next section.

3.3   Fixed points, equilibrium and stability

A point $\mathbf{r}_k$ where the phase velocity $v(\mathbf{r}_k)$ is zero is named a *fixed point* and represents a system in equilibrium:

$$v(\mathbf{r}_k) = 0 \quad \Longleftrightarrow \quad \mathbf{r}_k \text{ is a fixed point.} \tag{3.15}$$

Thus for a pendulum constrained to move in a vertical plane, the fixed points are the points in phase space which represent the pendulum when it points vertically downwards, and when it points vertically upwards.

A system in equilibrium stays in that state, but there is a negligible chance of it being precisely in equilibrium; its behaviour is then determined by the stability of the fixed point, and systems of higher order are more subtle than systems of first order. In our example of a pendulum the stability is clearly different for the two fixed points.

The fixed point $\mathbf{r}_k$ is said to be an *attractor* of some motion $\mathbf{r}(t)$ if

$$\lim_{t \to \infty} \mathbf{r}(t) = \mathbf{r}_k. \tag{3.16a}$$

(3.16*b*)

Attractor

The point $\mathbf{r}_k$ is *strongly* or *asymptotically stable* if it is an attractor for the motion along all the phase curves passing through some neighbourhood of $\mathbf{r}_k$.

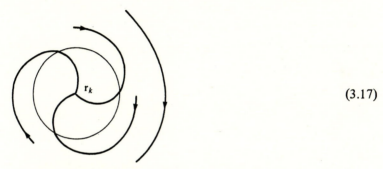

(3.17)

Strongly stable fixed point

It is often helpful to be specific by choosing as a neighbourhood the interior of a circle centred at $\mathbf{r}_k$.

Not all stability is strong stability. For the definition of stability in general we need the following: a region R_1 of phase space is said to *contain* the motion from some other region of phase space R_2, if all motion originating in R_2 at $t = 0$ remains within R_1 for $t \geqslant 0$, that is

$$\mathbf{r}(0) \in R_2 \Rightarrow \mathbf{r}(t) \in R_1 \quad (t \geqslant 0) \,. \tag{3.18a}$$

(3.18*b*)

Clearly R_2 is entirely within R_1.

A fixed point $\mathbf{r}_k$ is *stable* if every neighbourhood of $\mathbf{r}_k$ contains the motion from some other neighbourhood of $\mathbf{r}_k$. We have illustrated in (3.18*b*) the flow around a fixed point that is stable but not strongly stable. Such stability is important for Hamiltonian systems. A fixed·point that is not stable is *unstable*. All strongly stable points are stable, but not the other way round.

3.4   Separation of variables

There is a rich variety of fixed points. We consider only isolated fixed
points of second-order systems and introduce some particularly simple examples
using the method of separation of variables, by which a second-order system is
reduced to a pair of independent first-order systems. These simple examples then
form a framework upon which we later build the general theory of classification.

A second-order system is separable if it possesses coordinates (x_1, x_2) such
that the equations of motion have the form

$$\dot{x}_1 = v_1(x_1), \quad \dot{x}_2 = v_2(x_2). \tag{3.19}$$

We consider systems which are separable in Cartesian coordinates and others
which are separable in polar coordinates.

Clearly if $x_1 = a$ is a fixed point of one first-order system and $x_2 = b$ is
a fixed point of the other, then

$$\mathbf{r}_1 = (a, b) \tag{3.20}$$

is a fixed point of the combined second-order system. Furthermore the flow in
the neighbourhood of $\mathbf{r}_1$ will be determined by the flows in the neighbourhoods
of $x_1 = a$ and $x_2 = b$.

We can now use the theory of fixed points of first-order systems given in
chapter 1 to produce examples for second-order systems. It will help the reader
to refer back to the first chapter for these examples.

All our examples are linear, since they are the simplest, but they can also be
used to classify the fixed points of non-linear systems, as we show later.

Example 3.2 Cartesian separability
We combine two systems of the type in example 1.3 to obtain the equations of
motion

$$\dot{x} = \lambda_1 x, \quad \dot{y} = \lambda_2 y, \tag{3.21}$$

with solutions

$$x = x_0 e^{\lambda_1 t}, \quad y = y_0 e^{\lambda_2 t}. \tag{3.22}$$

The phase curves for the second-order system are obtained by eliminating t
from these equations to give

$$(x/x_0)^{\lambda_2} = (y/y_0)^{\lambda_1}. \tag{3.23}$$

The shapes of these phase curves depend upon the stability of equilibrium in
each sub-system. If they are both stable or both unstable λ_1/λ_2 is positive and
the phase curves are generalized parabolas passing through the origin, looking
like this *stable node* for $\lambda_2 < \lambda_1 < 0$

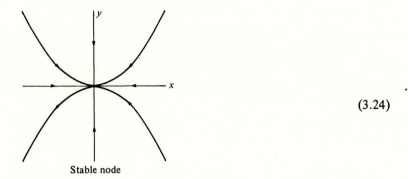

Stable node

(3.24)

or like this *unstable node* for $\lambda_2 > \lambda_1 > 0$:

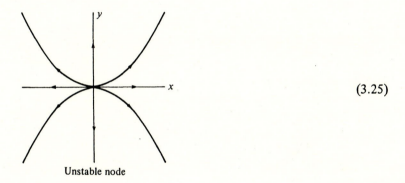

Unstable node

(3.25)

or like this *stable star* for $\lambda_2 = \lambda_1 < 0$:

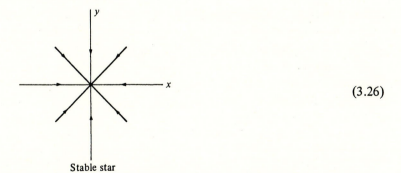

Stable star

(3.26)

All stable nodes are strongly stable. If $\lambda_1 < \lambda_2 < 0$ the fixed point is a stable node looking like (3.24), but rotated through a right angle; similarly for the unstable node with $\lambda_1 > \lambda_2 > 0$. The unstable star is produced when $\lambda_1 = \lambda_2 > 0$

and is obtained from the stable star by reversing the arrows.

If one of the sub-systems is stable and the other unstable, so that λ_1/λ_2 is negative, the phase curves are generalized hyperbolas looking like this *hyperbolic point* with $\lambda_1 < 0 < \lambda_2$,

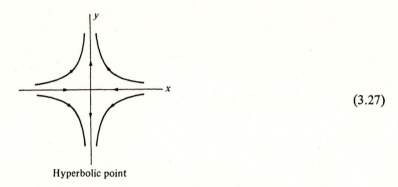

(3.27)

Hyperbolic point

with the arrows reversed if $\lambda_2 < 0 < \lambda_1$. Hyperbolic points are unstable.

Example 3.3 Polar separability

Suppose a linear system is separable in polar coordinates

$$x_1 = r, \quad x_2 = \theta.$$ (3.28)

Then the phase space of the first-order system in r has a natural boundary at $r = 0$ that must be a fixed point, and the phase space of the first-order system in θ must be a circle.

For linearity the equations of motion are

$$\dot{r} = \alpha r, \quad \dot{\theta} = \omega,$$ (3.29)

where α and ω are real constants, giving a fixed point at the origin of r-space. The first-order systems represented by equations (3.29) are presented as examples 1.3 and 1.2 of chapter 1.

By combining these examples, we see that the phase diagrams look like this *spiral·point* for $\alpha < 0, \omega > 0$:

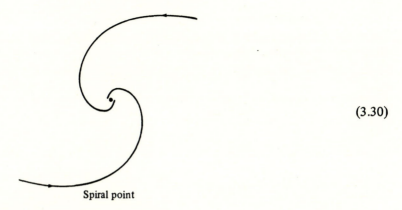

(3.30)

Spiral point

and this *spiral point* for $\alpha < 0$, $\omega < 0$.

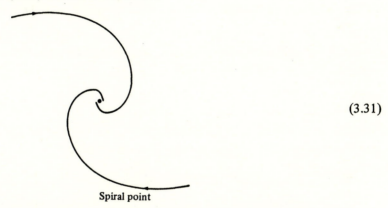

(3.31)

Spiral point

These are *strongly stable spiral points*. Reversing the signs of both α and ω reverses the arrows on the diagrams, giving *unstable spiral points*.

The solutions of equations (3.29) are

$$r = r_0 e^{\alpha t}, \quad \theta = \omega t + \theta_0, \tag{3.32}$$

so the spiral phase curves have equations

$$r = r_0 \exp [\alpha(\theta - \theta_0)/\omega]. \tag{3.33}$$

If $\omega = 0$ we again obtain stars; but for $\alpha = 0$ and $\omega > 0$ the phase diagram looks like this:

(3.34)

Elliptic point

This is a stable *elliptic point* or *centre* in which the ellipses happen to be circles. Elliptic points are stable, but not strongly stable. Changing the sign of ω reverses the arrows and preserves the stability.

In orthogonal Cartesian coordinates the equations of motion (3.29) have the form

$$\dot{\mathbf{r}} = A\mathbf{r}, \quad A = \begin{pmatrix} \alpha & -\omega \\ \omega & \alpha \end{pmatrix} \tag{3.35}$$

which is needed for the general treatment of the next section.

3.5 Classification of fixed points

The general theory is largely based on the above particular examples. Suppose, for simplicity, that the origin has been moved to the fixed point whose stability is being investigated and that the velocity function can be expressed in the form

$$v(\mathbf{r}) = A\mathbf{r} + O(|\mathbf{r}|^2) \tag{3.36}$$

where A is the non-singular, real, 2×2 matrix,

$$A = \begin{pmatrix} a & b \\ c & d \end{pmatrix}. \tag{3.37}$$

Sufficiently close to the origin the non-linear terms in $\mathbf{r}$ may be neglected, so that the stability can be analysed in terms of the linear motion:

$$\dot{\mathbf{r}} = A\mathbf{r}, \tag{3.38}$$

or

$$\left. \begin{aligned} \dot{x} &= ax + by, \\ \dot{y} &= cx + dy. \end{aligned} \right\} \tag{3.39}$$

Let $\mathbf{R} = (X, Y)$ be new vector coordinates obtained from $\mathbf{r}$ by an invertible linear transformation,

$$\mathbf{R} = M\mathbf{r}, \tag{3.40}$$

where M is a constant non-singular 2×2 matrix. In the new coordinate system the equations of motion are

$$\dot{\mathbf{R}} = B\mathbf{R}, \quad \text{where} \quad B = MAM^{-1}. \tag{3.41}$$

According to the general theory of chapter 2, for any matrix A, a transformation matrix can be found for which B is type 1, 2 or 3. Each of these gives rise to a different type of fixed point.

Type 1: Eigenvalues, λ_1 *and* λ_2*, real and distinct.*
The equations of motion in the (X, Y) coordinates are

$$\dot{X} = \lambda_1 X, \quad \dot{Y} = \lambda_2 Y, \tag{3.42}$$

which have the same form as equations (3.21) and the fixed point is a stable or an unstable node, a star, or a hyperbolic point, as shown in example 3.2, for various conditions on λ_1 and λ_2.

Type 2: Complex eigenvalues $\lambda_1 = \alpha + i\omega$, $\lambda_2 = \alpha - i\omega$.
The equations of motion in the (X, Y) coordinates have the form (3.35) and the fixed point is a stable or unstable spiral point or an elliptic point, as shown in example (3.3).

Type 3: Eigenvalues real and equal.
There are two separate cases to consider. If $b = c = 0$ then $a = d$; the equations of motion are given by (3.21) and the fixed point is a stable or unstable star.

If $c \neq 0$, then, from chapter 2

$$B = \begin{pmatrix} \lambda & 0 \\ c & \lambda \end{pmatrix} \tag{3.43}$$

and the equations of motion (3.41) and their solutions become

$$\dot{X} = \lambda X, \qquad \dot{Y} = cX + \lambda Y;$$
$$X = X_0 e^{\lambda t}, \qquad Y = (Y_0 + cX_0 t) e^{\lambda t}. \tag{3.44}$$

For sufficiently large t, the term Y_0 can be neglected and we can eliminate t to obtain

$$Y/X = \frac{c}{\lambda} \ln (X/X_0). \tag{3.45}$$

Suppose $\lambda < 0$, then $\lambda^{-1} \ln (X/X_0)$ is negative for sufficiently early times and positive for sufficiently late times, and since X retains its sign, Y changes sign. Also it follows from equations (3.44) that $(Y/X) \to \infty$ as $Y \to 0$ and the phase curves are tangential to the Y-axis, giving the following diagram for what is named a stable *improper node*, with $\lambda_1 = \lambda_2 < 0$ and $b \neq 0$ or $c \neq 0$.

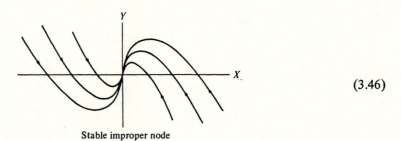

Stable improper node

(3.46)

If $\lambda > 0$ we obtain an unstable improper node, with the arrows reversed.

This completes our treatment of fixed points using the coordinates $\mathbf{R} = (X, Y)$. The type of equilibrium is unchanged in the original coordinate system $\mathbf{r} = (x, y)$, but the shape is changed by the transformation $\mathbf{r} = M^{-1}\mathbf{R}$, so there is a much greater variety of flows around fixed points. We illustrate some of these in the following summary.

3.6 Summary of classification

The various types of fixed points are given below.

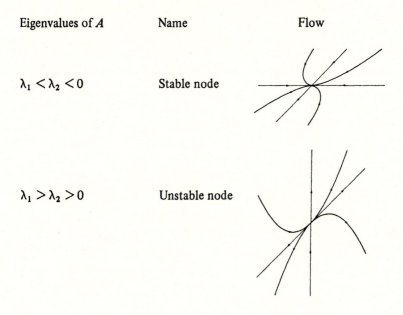

Eigenvalues of A	Name	Flow
$\lambda_1 < \lambda_2 < 0$	Stable node	
$\lambda_1 > \lambda_2 > 0$	Unstable node	

$\lambda_1 < 0 < \lambda_2$ Hyperbolic point (unstable)

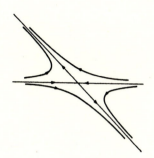

$\lambda_1 = \lambda_2 < 0$
$b = c = 0$ Stable star

$\lambda_1 = \lambda_2 > 0$
$b = c = 0$ Unstable star

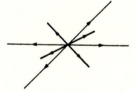

$\lambda_1 = \lambda_2^* = \alpha + i\omega$
$\alpha < 0, \omega < 0$ Stable spiral point

$\lambda_1 = \lambda_2^* = \alpha + i\omega$ Unstable spiral point

$\alpha > 0,\ \omega < 0$

$\lambda_1 = \lambda_2^* = i\omega$ Elliptic point

$\lambda_1 = \lambda_2 < 0$

$b \neq 0$ or $c \neq 0$ Stable

                            improper node

The unstable improper node with $\lambda_1 = \lambda_2 > 0$ has a similar diagram with the arrows reversed. We do not consider those cases in which eigenvalues are zero.

3.7   Determination of fixed points

When we know the velocity function $v(x, y)$ of a second-order system we can understand many features of the behaviour of the system from a study of the phase flow. In the phase diagram each fixed point shows us the structure of the phase flow in its neighbourhood and together they often give us a good indication of the nature of the flow in the whole of phase space.

For a given system we first solve the simultaneous equations

$$v_x(x, y) = 0, \quad v_y(x, y) = 0 \tag{3.47}$$

to determine the position of the fixed points $\mathbf{r}_k$, and then obtain the A-matrix for the *linearized* flow in the neighbourhood of each $\mathbf{r}_k$. This is done using a Taylor expansion of the velocity function around each $\mathbf{r}_k$. The first non-zero term in the expansion is the linear term so

$$\begin{pmatrix} v_x(x, y) \\ v_y(x, y) \end{pmatrix} = \begin{pmatrix} \partial v_x/\partial x & \partial v_x/\partial y \\ \partial v_y/\partial x & \partial v_y/\partial y \end{pmatrix} \begin{pmatrix} x - x_k \\ y - y_k \end{pmatrix} + O(\mathbf{r} - \mathbf{r}_k)^2, \quad (3.48)$$

where the derivatives are evaluated at $\mathbf{r}_k$. Therefore, around $\mathbf{r}_k$,

$$v = A(\mathbf{r} - \mathbf{r}_k) + O(\mathbf{r} - \mathbf{r}_k)^2, \quad (3.49)$$

where

$$A = \begin{pmatrix} \partial v_x/\partial x & \partial v_x/\partial y \\ \partial v_y/\partial x & \partial v_y/\partial y \end{pmatrix}. \quad (3.50)$$

In practice we evaluate this matrix as a function of x and y, and then substitute the values at the fixed points.

Example 3.4
Find the fixed points of a system whose coordinate x satisfies the differential equation

$$\ddot{x} - \dot{x}^2 + x^2 - x = 0 \quad (3.51)$$

and determine their type and their stability.

The equation is of second order, so to put it into standard form we write

$$v_x = \dot{x} = y \quad (3.52)$$

so that $v_y = \dot{y} = y^2 - x^2 + x$.

The fixed points are at the roots of $v_x = 0 = v_y$, that is the origin $(0, 0)$ and the point $(1, 0)$. The A-matrix is

$$\begin{pmatrix} \partial v_x/\partial x & \partial v_x/\partial y \\ \partial v_y/\partial x & \partial v_y/\partial y \end{pmatrix} = \begin{pmatrix} 0 & 1 \\ 1 - 2x & 2y \end{pmatrix}. \quad (3.53)$$

At the origin, $A = \begin{pmatrix} 0 & 1 \\ 1 & 0 \end{pmatrix}$ and the eigenvalues are $\lambda = \pm 1$. They are real, so the origin is a hyperbolic fixed point, which is unstable. At the point $(1, 0)$, $A = \begin{pmatrix} 0 & 1 \\ -1 & 0 \end{pmatrix}$ and the eigenvalues are $\lambda = \pm i$ so this is an elliptic fixed point, which is stable, but not strongly stable.

3.8   Limit cycles
For first-order systems all motions tend to fixed points or to infinity. This is not the case for second-order systems. Consider a system separable in polar coordinates, where the radial motion has a fixed point other than the origin, for example

$$\dot{r} = \alpha r(r - R) \quad (R > 0),$$
$$\dot{\theta} = \omega. \quad (3.54)$$

Clearly the circle $r = R$ is invariant, the phase points cycle around it indefinitely, so it is known as a cycle. In the neighbourhood of $r = R$ the radial motion has the linear form

$$\dot{s} = \lambda s \quad (s = r - R, \ \lambda = \alpha R), \tag{3.55}$$

so that the motion is given by

$$r - R = (r_0 - R) e^{\lambda t}, \quad \theta = \omega t + \theta_0 \tag{3.56}$$

and the phase curves by

$$r = R + (r_0 - R) \exp [\lambda(\theta - \theta_0)/\omega]. \tag{3.57}$$

For $\lambda < 0$ motion in the neighbourhood of the cycle approaches it as $t \to \infty$, as shown in the phase diagram.

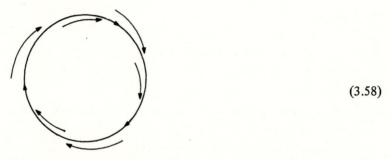

$$\tag{3.58}$$

Limit cycle, attractor

The attractor is a strongly stable invariant set. If λ is positive, then the arrows point away from the cycle. Systems tend away from such cycles, which are unstable.

Limit cycles are usually structurally stable, and are very important for the theory of oscillators, but we do not consider them in detail.

Suppose in equations (3.54) that $\alpha = 0$. In that case all circles centred on the origin are cycles, but none is a limit cycle. This case will be considered in more detail when we deal with Hamiltonian systems.

Exercises for chapter 3

(1) Sketch the phase diagram for systems with the following velocity functions, where a and b are constants with $b > a > 0$:

(a) $v(\mathbf{r}) = (a, b)$;　　　　　　　(c) $v(\mathbf{r}) = (x^2, y)$;

(b) $v(\mathbf{r}) = (a, x)$;　　　　　　　(d) $v(\mathbf{r}) = (2xy, y^2 - x^2)$.

(2) Classify the fixed points of the following linear systems and state whether they are unstable, stable or strongly stable:

(a) $\dot{x} = 3x + 4y$, $\dot{y} = 2x + y$;
(b) $\dot{x} = 3x$, $\dot{y} = 2x + y$;
(c) $\dot{x} = x + 2y$, $\dot{y} = -2x + 5y$;
(d) $\dot{x} = x + 2y$, $\dot{y} = -5x + 4y$.

(3) Classify the fixed points and discuss the stability for the following linear systems: .

(a) $\dot{x} = x + 3y$, $\dot{y} = -6x + 5y$;
(b) $\dot{x} = x + 3y + 4$, $\dot{y} = -6x + 5y - 1$;
(c) $\dot{x} = x + 3y + 1$, $\dot{y} = -6x + 5y$;
(d) $\dot{x} = 3x + y + 1$, $\dot{y} = -x + y - 6$.

(4) For each of the following systems find the isolated fixed points and classify them:

(a) $\dot{x} = -4y + 2xy - 8$, $\dot{y} = 4y^2 - x^2$;
(b) $\dot{x} = y - x^2 + 2$, $\dot{y} = 2(x^2 - y^2)$.

(5) The motion of a damped linear oscillator is described by the equation

$$\ddot{x} + \alpha\dot{x} + \omega^2 x = 0, \quad (0 < \alpha < 2\omega) .$$

Express the equation in standard form and obtain the phase velocity function. Where is the fixed point and what is its type? Sketch the phase diagram. Describe what happens to the motion and to the fixed point as $\alpha \to 0$ and when $\alpha = 0$.

(6) Write the following second-order equations in standard form, find their isolated fixed points and classify them:

(a) $\ddot{x} - \dot{x} + x^2 - 2x = 0$;
(b) $\ddot{x} - \dot{x}^3 + x + 5 = 0$.

(7) The population of a host, $H(t)$, and a parasite, $P(t)$, are described approximately by the equations

$$\frac{dH}{dt} = (a - bP)H, \quad \frac{dP}{dt} = (c - \frac{dP}{H})P, \quad H > 0 ,$$

where a, b, c and d are positive constants. By a suitable change of scales show that these equations may be put in the simpler form

$$\frac{dy}{d\tau} = (1-x)y, \quad \frac{dx}{d\tau} = \alpha x(1 - \frac{x}{y}) ,$$

where P, H and t are proportional to x, y and τ respectively and $\alpha = c/a$. Find and classify the fixed point of these equations. In the case $\alpha = \frac{1}{2}$ sketch the phase flow across the following lines:

(a) $y = x$;   (c) $x = 0$;
(b) $x = 1$;   (d) $y = \beta x$.

In the last case choose values of the constant β greater and less than unity. Using this information sketch the whole phase diagram.

(8) A population F of foxes feeds on a population H of hares. The total birth rate of the foxes is proportional to the population of foxes and to the amount of food available, that is, to FH. Hares die through encounters with foxes and their total birth rate is proportional to the population of hares. Foxes die at a rate proportional to their population. With these assumptions show that the rates of change of population obey the equations

$$\frac{dH}{dt} = aH - bHF \quad (H > 0),$$

$$\frac{dF}{dt} = cHF - dF \quad (F > 0),$$

where a, b, c and d are positive constants that you should define. Find the fixed point of the system and describe the motion in its neighbourhood. What does this mean in terms of the populations of hares and foxes?

(9) In exercise 8 show that all phase curves are closed, and sketch a typical graph of $F(t)$. If the population of hares is suddenly depleted by an epidemic disease from which the remaining hares are immune discuss qualitatively the various different effects this could have on the system.

(10) A damped vertical pendulum has the equation of motion

$$\ddot{\psi} + \alpha\dot{\psi} + \omega^2 \sin \psi = 0 \quad (\alpha > 0),$$

where ψ is the angle between the pendulum and the downward vertical. Determine the position and nature of the fixed points. Sketch the phase curves in a sufficiently large region of phase space to show all the qualitative features of the motion. Describe what happens to the motion and fixed points as $\alpha \to 0$ and when $\alpha = 0$.

(11) Find the limit cycles for the system:

$$\dot{x} = y + x(1 - x^2 - y^2),$$
$$\dot{y} = -x + y(1 - x^2 - y^2).$$

(12) Classify all the fixed points and limit cycles (if any) of the two-dimensional flow defined by the equation

$$\dot{\mathbf{r}} = \mathbf{r}\,(a \sin br^2)/r^2 + c(x\hat{\mathbf{y}} - y\hat{\mathbf{x}}),$$

where a, b and c are positive constants and

$$\mathbf{r} = x\hat{\mathbf{x}} + y\hat{\mathbf{y}} \qquad r = |\mathbf{r}| \ .$$

(13) A boat is rowed across a river of width a; it is rowed so that it always points towards a fixed point on the opposite bank. The boat moves at a constant speed u relative to the water which flows at a constant speed v. Show that

$$\dot{x} = -ux/r, \quad \dot{y} = v - uy/r, \quad r^2 = x^2 + y^2,$$

where the y-axis is taken to be one bank and the origin the point towards which the boat points: the line $x = a$ is the other bank. Show that the phase curves are given by

$$r + y = a^\alpha x^{1-\alpha} \quad (\alpha = v/u).$$

Sketch the phase curves for $\alpha < 1$ in the neighbourhood of the origin. What happens if $\alpha > 1$?

(14) Show that the origin is an elliptic fixed point for the system described by the equations

$$\dot{x} = -y + xr^2 \sin(\pi/r), \quad \dot{y} = x + yr^2 \sin(\pi/r), \quad r^2 = x^2 + y^2.$$

Express the equations in polar coordinates to show

(a) that the circles $r = 1/n, n = 1, 2, \ldots$ are phase curves;

(b) that the trajectories between any two consecutive circles spiral either away from or towards the origin;

(c) that the phase curves outside $r = 1$ are unbounded.

4 CONSERVATIVE HAMILTONIAN SYSTEMS OF ONE DEGREE OF FREEDOM

4.1 Newtonian and Hamiltonian systems

The original dynamical systems were *Newtonian* and the classical example of Newton's work was his treatment of the solar system as a collection of bodies moving according to his basic dynamical laws with a definite force, given in this case by the inverse square law. Such a system is also *Hamiltonian* in the sense which we define in this section. Of course it has very many degrees of freedom, whereas we shall be concerned with only one. Most of the systems of chapters 1 and 3 were not Hamiltonian. The Hamiltonian systems form a limited but important class of dynamical systems, the specialized theory of which is introduced in this chapter.

The *configuration* of a dynamical system of one degree of freedom is represented by a point in a one-dimensional configuration space with coordinate q. Examples are the height of a particle constrained to move vertically, or the angle of rotation of a pendulum whose motion is confined to a vertical plane.

Suppose a dynamical system obeys Newton's laws of motion: then the configuration at time t does not determine the motion for all time; a momentum p is also required. The coordinate q and momentum p define in the two-dimensional phase space a point (q, p) representing a state of the system. The state at time t does determine the motion for all time. A Newtonian system of one degree of freedom is a dynamical system of second order.

Let q be the displacement in a given direction of a particle of mass m, let p be its linear momentum in that direction, and suppose the corresponding force on the particle is given by the function $F(q, t)$. The equation of motion and the definition of momentum are then given in standard form (3.3) by

$$\dot{q} = p/m \tag{4.1a}$$

$$\dot{p} = F(q, t). \tag{4.1b}$$

The motion of the system is given by the solution of these equations, which have the standard form (3.1) for second-order systems with

$$\mathbf{r} = (q, p), \quad v(q, p, t) = (p/m, F(q, t)), \tag{4.2}$$

where v is the phase velocity. Remember this is not the same as the velocity $\dot{q}$ of the particle in ordinary configuration space.

Because the system has only one degree of freedom, the force in equation $(4.1b)$ can always be derived from a potential $V(q, t)$:

$$F(q, t) = -\frac{\partial V}{\partial q} (q, t), \tag{4.3}$$

where

$$V(q, t) = -\int_{q_0}^{q} dq' \, F(q', t). \tag{4.4}$$

Similarly, the right-hand side of $(4.1a)$ can be expressed as a derivative

$$p/m = \frac{d}{dp} \left(\frac{p^2}{2m} \right). \tag{4.5}$$

We define the *Hamiltonian function* or *Hamiltonian* of the system to be

$$H(q, p, t) = \frac{p^2}{2m} + V(q, t), \tag{4.6}$$

so that the equations (4.1) can be rewritten as *Hamilton's equation* of motion, which are

$$\dot{q} = \frac{\partial H}{\partial p} (q, p, t) \tag{4.7a}$$

$$\dot{p} = -\frac{\partial H}{\partial q} (q, p, t) \tag{4.7b}$$

and the phase velocity is

$$v = \left(\frac{\partial H}{\partial p}, \frac{-\partial H}{\partial q} \right). \tag{4.8}$$

In the example given the Hamiltonian (4.6) is the sum of a kinetic energy, $p^2/2m$, and a potential energy, but there are many systems in which the Hamiltonian does not have this form.

A *Hamiltonian system* of one degree of freedom is a second-order system whose motion is determined by equations of the form (4.7). Besides the well-known mechanical systems, there are electrical, biological, meteorological and economic Hamiltonian systems.

For all cases the coordinate q in equation (4.7) is known as a *generalized coordinate* and the quantity p as its *conjugate momentum*: the pair (q, p) are called *conjugate variables*. In general q *need* not represent the configuration of the system nor *need* p be a physical momentum, although they frequently do have this meaning. It is because of this generality that we use q instead of x for the displacement of a particle, changing notation from that used in Chapter 3.

4.2 Conservative systems

An autonomous Hamiltonian system has a Hamiltonian function

$$H = H(q,p) \tag{4.9}$$

that is independent of time, and is named a *conservative system*.

For a given motion, using Hamilton's equations (4.7), the rate of change of the value of the Hamiltonian is given by

$$\frac{dH}{dt} = \frac{\partial H}{\partial q}\dot{q} + \frac{\partial H}{\partial p}\dot{p}$$

$$= \frac{\partial H}{\partial q}\frac{\partial H}{\partial p} + \frac{\partial H}{\partial p}\left(-\frac{\partial H}{\partial q}\right) = 0, \tag{4.10}$$

so that the value of the Hamiltonian is always conserved, which is why the system is named conservative. For Newtonian systems this value is equal to the energy E.

The gradient of the Hamiltonian function is

$$\nabla H(q,p) = \left(\frac{\partial H}{\partial q}, \frac{\partial H}{\partial p}\right). \tag{4.11}$$

From (4.10) and (4.11) we see that the phase velocity $v = (\dot{q},\dot{p})$ of a conservative system is equal in magnitude and perpendicular in direction to the gradient of the Hamiltonian.

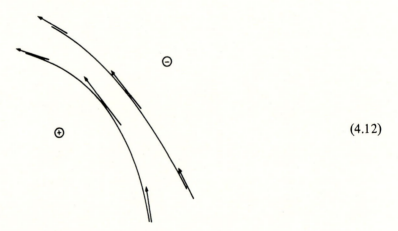

$$\tag{4.12}$$

Since the Hamiltonian is conserved the motion is along curves of constant H. Because of these properties the phase diagram of a Hamiltonian system is drawn with contours, or curves of constant H, the difference between the values of H on neighbouring contours being the same throughout the phase plane. This is analogous to a geographical contour map, in which $H(q, p)$ represents the height above sea level of the point (q, p). From (4.8) it follows that the magnitude of the phase velocity is inversely proportional to the distance between neighbouring contours. The 'high' and 'low' ground can be indicated by $\oplus$ and $\odot$ signs. In that case the phase curves, which are also contours, are followed by keeping the $\oplus$ signs to the left and the $\odot$ signs to the right. All information about the motion of a conservative Hamiltonian system is provided by this simplified phase diagram and there is no need to draw a mesh of arrows on the plane as for the general autonomous systems of chapters 1 and 3. Examples are given in the following sections.

The contours of the Hamiltonian function $H(q, p)$ are invariant sets of the system. Hamiltonian systems have fixed points which are also invariant sets. The fixed points occur where $v = 0$, so that by equation (4.8) the gradient of the Hamiltonian is zero:

$$\left(\frac{\partial H}{\partial q} , \frac{\partial H}{\partial p} \right) = \nabla H = 0 \quad \text{(fixed point).} \tag{4.13}$$

At the fixed points the system is in equilibrium. Their location and nature helps us to determine the overall pattern of the motion. This is clearly illustrated in the following section on linear systems.

4.3 Linear conservative systems

As in the previous chapter we first present some simple examples of linear behaviour, and then show that these examples can be used for a general analysis. We take our examples from the Newtonian mechanics of a particle of mass m moving in a time-independent potential $V(q)$. The system is then conservative with Hamiltonian

$$H(q, p) = \frac{p^2}{2m} + V(q), \tag{4.14}$$

from which we obtain Newton's equation of motion

$$m\ddot{q} = -\mathrm{d}V/\mathrm{d}q = F(q), \tag{4.15}$$

where $F(q)$ is the force.

For the equations of motion to be linear the potential $V(q)$ must be at most quadratic.

Example 4.1 The uniform force field
The potential is

$$V(q) = aq \quad (a > 0) \tag{4.16a}$$

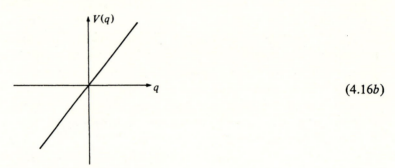

$$\tag{4.16b}$$

and the energy contours are parabolas

$$aq = E - p^2/2m \tag{4.17a}$$

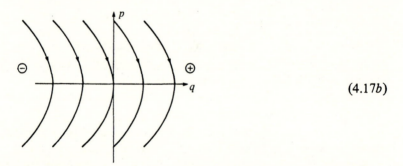

$$\tag{4.17b}$$

There are no fixed points. The example 3.1 of the previous chapter is a special case. Here we are using the more effective methods which can only be used for Hamiltonian systems.

The motion is obtained by solving equation (4.15) to give

$$q = q_0 - \frac{a}{2m}(t - t_0)^2, \quad p = -a(t - t_0), \tag{4.18}$$

where t_0 is the time at which $p = 0, q = q_0$.

Example 4.2 The linear oscillator
This is also known as the simple harmonic oscillator. The force is linear in q and directed towards the origin, so the potential is quadratic:

$$V(q) = \tfrac{1}{2}aq^2 \quad (a > 0). \tag{4.19a}$$

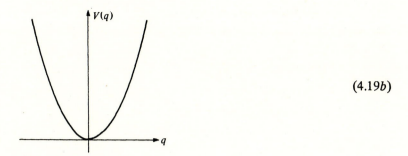

(4.19*b*)

The energy contours are the ellipses

$$p^2 + m^2 \omega^2 q^2 = 2mE \quad (\omega^2 = a/m),$$ (4.20*a*)

centred at the fixed point $q = p = 0$, which is clearly an elliptic point, as in example 3.3 and section 3.6.

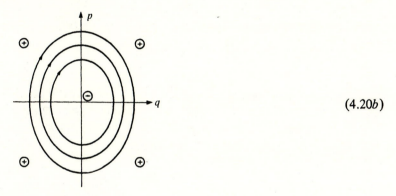

(4.20*b*)

The equation of motion is

$$\ddot{q} = -\omega^2 q$$ (4.21)

with solution

$$q = A \cos(\omega t + \delta), \quad p = -m\omega A \sin(\omega t + \delta),$$ (4.22)

so the motion is periodic with period,

$$T = 2\pi/\omega,$$ (4.23)

which is independent of the amplitude A, a peculiarity of linear systems. The energy and amplitude are related by

$$E = \tfrac{1}{2} m A^2 \omega^2 .$$ (4.24)

The particle is said to oscillate, or librate, around the fixed point $q = p = 0$.

Example 4.3 The linear repulsive force.
The force is linear and away from the origin so the potential is a quadratic
barrier

$$V(q) = -\tfrac{1}{2} aq^2 \quad (a > 0).$$ (4.25a)

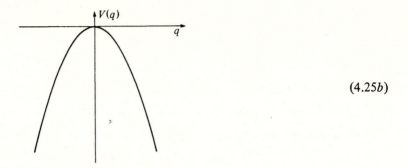

(4.25b)

The energy contours are the hyperbolas:

$$2mE = p^2 - m^2 \gamma^2 q^2 \quad (\gamma^2 = a/m),$$
$$= (p + m\gamma q)(p - m\gamma q).$$ (4.26)

There is one fixed point at the origin, $q = p = 0$ and the energy contours
consist of the straight lines $p = \pm m\gamma q$ through this fixed point, together with
hyperbolas asymptotic to these lines: the fixed point is a hyperbolic point and
is clearly unstable, as in example 3.2 and section 3.6.

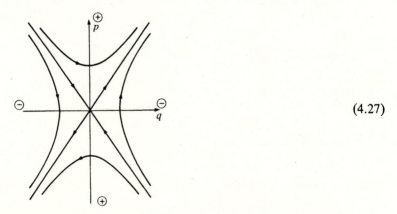

(4.27)

In general a phase curve which meets a hyperbolic point is known as a
separatrix. In this example the diagonal straight lines $p = \pm m\gamma q$ are *separatrixes*.
A separatrix marks a boundary between phase curves with different properties,
so it is a very important phase curve.

In our example the phase curves on the left of the straight lines represent motion towards the barrier from the left with subsequent reversal, because there is insufficient energy $(E < 0)$ to surmount the potential barrier. Similarly for the phase curves on the right. The phase curves at the top represent motion with sufficient energy $(E > 0)$ to cross the barrier from the left and the phase curves at the bottom represent motion across the barrier from the right.

The upper left separatrix represent an asymptotic approach, with $E = 0$, to the top of the barrier from the left, and the lower right separatrix represents an asymptotic approach from the right. The upper right and lower left separatrixes represent motion from near the top of the barrier to infinity in either direction.

The equation of motion is

$$\ddot{q} = \gamma^2 q \tag{4.28}$$

with solution

$$q = A_+ e^{\gamma t} + A_- e^{-\gamma t}, \quad p = m\gamma \left(A_+ e^{\gamma t} - A_- e^{-\gamma t}\right), \tag{4.29}$$

with energy

$$E = -2m\gamma^2 A_+ A_-. \tag{4.30}$$

The signs of A_+ and A_- determine the type of motion.

4.4 The cubic potential

We now describe the qualitative features of the motion of a particle of unit mass moving in the cubic potential

$$V(q) = \tfrac{1}{2} \omega^2 q^2 - \tfrac{1}{3} A q^3 \quad (A > 0). \tag{4.31}$$

The function $V(q)$ has a double zero at the origin and another zero at $q = 3\omega^2/2A$. It has stationary points at

$$0 = \partial V/\partial q = \omega^2 q - A q^2, \tag{4.32}$$

that is, at the origin and at $q = q_1 = \omega^2/A$. Near the origin, the cubic term is small by comparison with the quadratic term and the potential has the approximate form (4.19). It is a minimum. The point q_1 must therefore be a maximum, and the value of the potential there is

$$V(q_1) = \omega^6/6A^2. \tag{4.33}$$

The Hamiltonian of the system is

$$H(q, p) = \tfrac{1}{2} p^2 + \tfrac{1}{2} \omega^2 q^2 - \tfrac{1}{3} A q^3. \tag{4.34}$$

The fixed points are at $(0, 0)$ and $(0, q_1)$, the first being an elliptic fixed point as in example 4.2 and the second being a hyperbolic fixed point as in example 4.3, because the potential is a maximum at q_1. The separatrix is the energy contour which passes through the hyperbolic fixed point, so its energy is $E_s = V(q_1)$ and its equation is

$$\tfrac{1}{2}p^2 + \tfrac{1}{2}\omega^2 q^2 - \tfrac{1}{3}Aq^3 = \omega^6/6A^2. \tag{4.35}$$

This intersects the p-axis where $q = 0$ so $p = \pm\omega^3/A\sqrt{3}$. In drawing the phase diagram we start by drawing the potential function and then the separatrix. The rest of the diagram is then not difficult to sketch, and is shown in figure 4.1.

The motion depends on the energy. If it is greater than E_s the particle approaches from the right, is slowed by the potential barrier near q_1, accelerated again near the origin, and reflected by the potential barrier to the left of the origin, to return eventually to the right of the barrier, with increasing speed. For energy less than E_s the motion divides into two parts. The particle may oscillate about the potential minimum at the origin: the angular frequency of this oscillation approaches ω if the energy is close to zero. Or the particle may approach from the right and be reflected before reaching q_1.

Fig. 4.1 Cubic potential, equation (4.31) and phase diagram

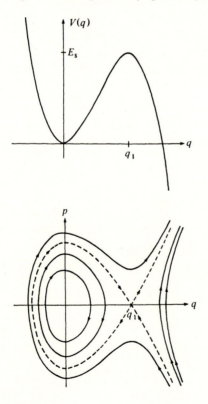

4.5   General potential

In general the phase diagram of a particle moving on the real line has both elliptic and hyperbolic fixed points, corresponding to the minima and the maxima of the potential function $V(q)$. Usually the motion in the neighbourhood of the fixed points resembles the corresponding linear motion described in the previous section. The hyperbolic points are joined by separatrixes that divide the phase space into invariant regions with different types of motion.

These features are illustrated by the example given in figure 4.2 in which a particle moves in the potential $V(q)$. The fixed points are at A, B, C and D. A and C are elliptic at the minima of the potential, whereas B and D are hyperbolic at the maxima of the potential.

The separatrix s_1 leading from B divides oscillatory motion in the potential well at A, for example the phase curve a, from the oscillatory motion in the two wells at A and C with sufficient energy to cross the potential barrier at B, for example the phase curve b. The separatrix s_2 leading from B divides motion like c in the well at C from motion like b.

The separatrix s_3 divides oscillatory motion like b in the potential wells from the non-oscillatory motion like d in which the particle approaches from infinity with sufficient energy to cross the potential barriers at B and D, but is reversed by the higher potential barrier to the left of A to return towards infinity on the right. The separatrixes s_4 divide motion like d from motion like e in which a particle from infinity on the right has insufficient energy to cross the barrier at D and is reversed by it to return towards infinity to the right.

In general, for a given motion, a point like E, F or G in configuration *or* phase space at which the momentum changes sign is known as a *turning point*.

The motion on the separatrix s_1 leaves B slowly to the left, crosses the well, and returns towards B, approaching B asymptotically, but never reaching the top of the potential barrier.

Any potential problem for one degree of freedom can be studied in this way, provided dV/dq exists and has only simple zeros. If it does not there may be complications.

In this example, and those of the previous section, the only types of isolated fixed points are elliptic and hyperbolic. None of the others appears. This is general and follows directly from the fact that the phase curves are contours of constant Hamiltonian. From Hamilton's equations, the Hamiltonian must be stationary at fixed points and there are only three types of structurally stable stationary point of real functions of two variables: maxima, minima and saddle points. The first two are elliptic points, although a maximum never occurs when the Hamiltonian has the form (4.14), and the last is a hyperbolic point. Equilibrium of Hamiltonian systems is never strongly stable.

Fig. 4.2 Complicated potential and phase diagram. For clarity the phase curves are not equally spaced in energy. Their energies are marked on the vertical axis of the top graph. The separatrixes s_1 and s_4 are phase curves that end at the hyperbolic fixed points corresponding to the maxima of the potential. E, F and G are turning points of the motion for the phase curves upon which they lie.

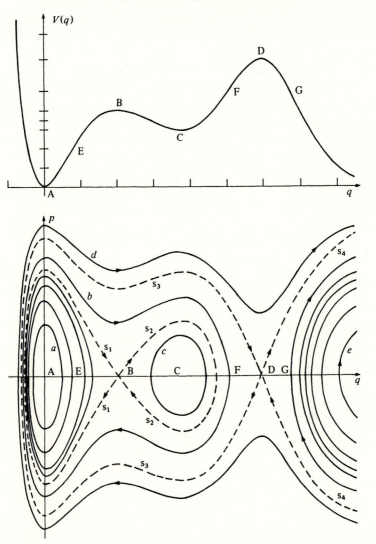

4.6   Free rotations

The configuration space does not always consist of the whole real line $(-\infty, \infty)$. For rotational motion the configuration space is a circle and it is convenient to choose a coordinate ψ in the range $[0, 2\pi]$ with the ends of the range identified. The Hamiltonian function, $H(\psi, p)$, is then periodic in ψ with period 2π.

For a body which rotates freely about a fixed axis, the Hamiltonian is a function of the angular momentum p only, and has the form

$$H(\psi, p) = p^2/2G \qquad \text{(free rotations),} \qquad (4.36)$$

where G is the moment of inertia which we shall put equal to 1.

The angular momentum p can take on any value in $(-\infty, \infty)$, depending upon the rate and direction of rotation, so the phase space may be represented on an infinitely long cylinder, as illustrated in figure 4.3.

Contours of constant H, or constant energy, are illustrated in figure 4.3, or unwrapped onto the plane of the paper as shown in (4.37), where the two sides of the rectangle represent the same line $\psi = \pi$.

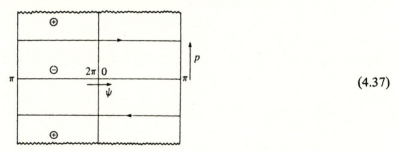

$$(4.37)$$

Fig. 4.3   Cylindrical representation of phase space for a freely rotating body, showing contours of constant H.

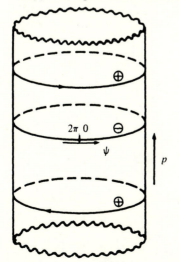

Hamilton's equations of motion,

$$\dot{\psi} = p, \quad \dot{p} = 0, \tag{4.38}$$

have the solution

$$\psi = pt + \psi_0, \quad p = p_0, \tag{4.39}$$

where (ψ_0, p_0) represents the state at time $t = 0$. Thus the line $p = 0$ consists of fixed points that are neither elliptic nor hyperbolic.

As might be expected from the presence of the non-isolated fixed points the system is not structurally stable. A small ψ-dependent change in the Hamiltonian completely changes the behaviour for small momenta, as shown by the next example.

4.7 The vertical pendulum

If a time-independent, or conservative, force is applied to the system of the previous section then the new Hamiltonian has the form

$$H(\psi, p) = p^2/2G + V(\psi), \tag{4.40}$$

where $V(\psi)$ is periodic in ψ with period 2π.

The simplest example is given by the vertical pendulum. A mass m at the point P is suspended from a fixed point O by a light rod of length l and constrained to move in a vertical plane through O in the presence of a uniform downward force F.

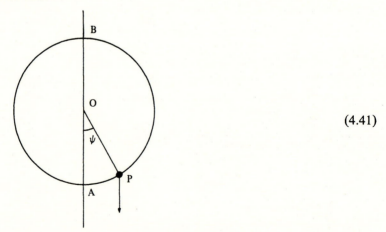

$$\tag{4.41}$$

Let ψ be the angle between the rod OP and the downward vertical OA. Then the Hamiltonian function is given by the energy:

$$H(\psi, p) = p^2/2G - Fl \cos \psi. \tag{4.42}$$

For simplicity we suppose that the moment of inertia $G = ml^2 = 1$. The general case may be obtained from this by rescaling. We define the constant α by $\alpha = (Fl)^{\frac{1}{2}}$, giving the Hamiltonian

$$H(\psi, p) = \tfrac{1}{2} p^2 - \alpha^2 \cos \psi. \tag{4.43}$$

A Hamiltonian of this form appears in the theory of particle motion in accelerators and plasmas, in the theory of molecular rotation, in some approximate theories of planetary motion, and in the general theory of systems of many degrees of freedom.

Fig. 4.4 Cylindrical phase space diagram and potential for vertical pendulum

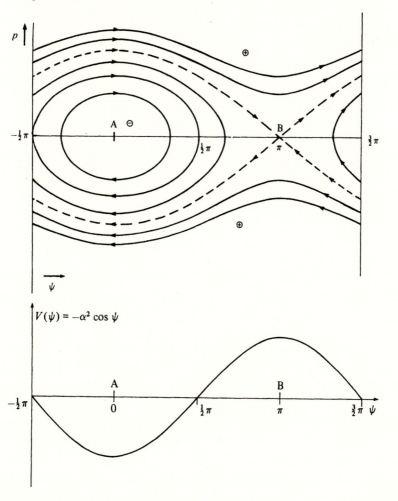

Hamilton's equations are

$$\dot\psi = p,$$

$$\dot p = -\alpha^2 \sin \psi,$$

(4.44)

which cannot generally be solved in terms of elementary functions. The cylindrical phase space diagram is shown in figure 4.4, divided at $\psi = \frac{3}{2}\pi$ to show both fixed points clearly. These are at

$$(\psi, p) = (0, 0), \quad (\psi, p) = (\pi, 0).$$

(4.45)

The fixed point at A is elliptic and corresponds to the minimum in the potential at $\psi = 0$, when the pendulum points vertically downwards. The fixed point at B is hyperbolic and corresponds to the maximum in the potential at $\psi = \pi$, when the pendulum is vertically upwards. The separatrix through B is the contour of energy α^2. Its equation can be obtained analytically, (see exercise 4.4) and it divides the phase space into three regions in each of which the motion is periodic. In the top and bottom regions the pendulum rotates clockwise or anti-clockwise, just as in the case of free rotation. But in the central region there is a new type of motion about the downward vertical. This new motion appears however small α might be.

4.8   Rotation, libration and periods

Clearly periodic motion can be divided into two types depending on whether or not the velocity changes sign during the course of the motion.

In general if the motion of a system is periodic and the coordinate q is expressed as a function of time, then the motion is named a

   rotation if $\dot q$ always has the same sign,   (4.46)

or a

   libration if $\dot q$ changes sign,   (4.47)

in the course of the motion. Oscillation and vibration are the same as libration. Clearly rotation cannot occur if the configuration space is the whole real line, as a continuously increasing or decreasing q cannot then give periodic motion. However, if the configuration space is a circle, either rotation or libration can occur. The types of periodic motion therefore depend upon the topology of the configuration space. Note that, when the phase space is represented on a plane, motion in a circle or any closed curve corresponds to libration. Rotation appears in the phase space as the graph of a periodic function of q.

In either case the time that the system takes to reach a given point can be obtained as an integral over the coordinate q if the Hamiltonian has the special form

$$H(q, p) = p^2/2m + V(q).$$

(4.48)

The momentum is given as a function of the energy E by

$$p = \pm \left[2m \left(E - V(q) \right) \right]^{1/2}, \tag{4.49}$$

so using Hamilton's first equation

$$\dot{q} = \pm \left[\frac{2}{m} \left(E - V(q) \right) \right]^{1/2} \tag{4.50}$$

and

$$\int^{t} dt' = \pm \left(\tfrac{1}{2} m \right)^{1/2} \int^{q} \frac{dq'}{[E - V(q')]^{1/2}}. \tag{4.51}$$

We can use this result to obtain the period T for rotation and libration. For rotation there are no turning points: if $q = \psi$ is the configuration angle, then

$$T = \int_{0}^{T} dt = \left(\tfrac{1}{2} m \right)^{1/2} \left| \int_{0}^{2\pi} \frac{d\psi}{[E - V(\psi)]^{1/2}} \right|. \tag{4.52}$$

For libration the period is twice the integral between the turning points q_1, q_2, which satisfy

$$V(q_1) = V(q_2) = E \quad (q_1 < q_2), \tag{4.53}$$

that is

$$T = (2m)^{1/2} \int_{q_1}^{q_2} \frac{dq}{[E - V(q)]^{1/2}}. \tag{4.54}$$

Gaussian numerical integration is a very effective method of obtaining the periods of libration in practice.

4.9 Area-preserving flows and Liouville's theorem

We introduced Hamiltonian systems in section 4.1 as a generalization of Newtonian systems, but they can be considered from quite a different point of view as those systems whose evolution preserves area in phase space. We need to show that systems which satisfy Hamilton's equations preserve the area. This is known as Liouville's theorem and, because it is so important we shall demonstrate it twice, once geometrically for conservative systems and once analytically for systems in which the Hamiltonian may be an explicit function of time.

For conservative systems we consider a small rectangular element of phase space between two neighbouring energy contours; then the phase velocity is inversely proportional to the local distance between contours.

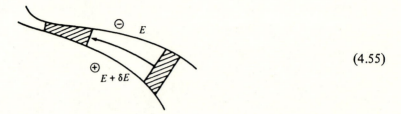

$$(4.55)$$

But the length of the element of phase space parallel to the energy contours is proportional to the phase velocity, so the area of the element is conserved. The area within any closed curve may be broken down into such small elements, and so the flow preserves area. In terms of fluids it is an incompressible flow.

A system which is influenced by time-dependent forces, or which is represented in a rotating or other non-inertial reference frame, has a Hamiltonian function $H(q, p, t)$ which depends explicitly upon the time. For a particular motion of such a system the rate of change in the value of the Hamiltonian with time is given by

$$\frac{\mathrm{d}H}{\mathrm{d}t} = \frac{\partial H}{\partial q}\,\dot{q} + \frac{\partial H}{\partial p}\,\dot{p} + \frac{\partial H}{\partial t} \tag{4.56}$$

and, on using Hamilton's equation

$$\frac{\mathrm{d}H}{\mathrm{d}t} = \frac{\partial H}{\partial q}\frac{\partial H}{\partial p} + \frac{\partial H}{\partial p}\left(\frac{-\partial H}{\partial q}\right) + \frac{\partial H}{\partial t}$$

$$= \frac{\partial H}{\partial t}. \tag{4.57}$$

Thus, at any time the variation of the value of H for the motion is given by the variation in H at the current point in the phase plane, considered as fixed. The value of the Hamiltonian is not conserved. The previous argument cannot therefore be used to demonstrate the preservation of area. However, the area *is* preserved, as we now show.

Let the time-dependent Hamiltonian be $H(q, p, t)$ so that Hamilton's equations are

$$\dot{q} = \frac{\partial H}{\partial p}, \quad \dot{p} = -\frac{\partial H}{\partial q}. \tag{4.58}$$

Let (q_0, p_0) and (q_1, p_1) be the phase points at times t_0 and t_1 for the motion along a particular phase curve.

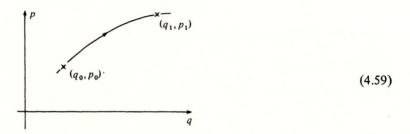

(4.59)

Then, for small times δt, where $t_1 = t_0 + \delta t$

$$q_1 = q(t_0 + \delta t) = q_0 + \delta t \frac{\partial}{\partial p_0} H(q_0, p_0, t_0) + O(\delta t^2)$$

$$p_1 = p(t_0 + \delta t) = p_0 - \delta t \frac{\partial}{\partial q_0} H(q_0, p_0, t_0) + O(\delta t^2).$$ (4.60)

Treating q_1 and p_1 as functions of (q_0, p_0) for fixed t_0 and δt, we find that the Jacobian of the transformation $(q_0, p_0) \rightarrow (q_1, p_1)$ is

$$\frac{\partial(q_1, p_1)}{\partial(q_0, p_0)} = \begin{vmatrix} \dfrac{\partial q_1}{\partial q_0} & \dfrac{\partial p_1}{\partial q_0} \\[2mm] \dfrac{\partial q_1}{\partial p_0} & \dfrac{\partial p_1}{\partial p_0} \end{vmatrix}$$

$$= \begin{vmatrix} 1 + \delta t \dfrac{\partial^2 H}{\partial q_0 \partial p_0} & -\delta t \dfrac{\partial^2 H}{\partial q_0^2} \\[3mm] \delta t \dfrac{\partial^2 H}{\partial p_0^2} & 1 - \delta t \dfrac{\partial^2 H}{\partial q_0 \partial p_0} \end{vmatrix} + O(\delta t^2)$$

$$= 1 + O(\delta t^2).$$ (4.61)

Thus the area enclosed by any closed curve changes by an amount proportional to at most δt^2. Any time interval T may be divided into N small intervals of duration $\delta t = T/N$, so that the total change in area is proportional to $N (T/N)^2$, which goes to zero as δt decreases.

Thus any flow satisfying Hamilton's equations is area-preserving. It may also be shown that if the velocity function of an area-preserving flow is sufficiently well-behaved, then the flow satisfies Hamilton's equations. Such a flow is named Hamiltonian.

This ends the similarity between time-independent and time-dependent Hamiltonians. The general theory of the latter is incomplete, and only for special classes of time-dependence does a general theory exist. We deal with some of these systems later in chapters 9, 10 and 11.

Exercises for chapter 4

(1) A Hamiltonian $H(q,p)$ is independent of q. Show that p remains constant and that q increases linearly with time. Sketch the phase diagram for

(a) $H(q,p) = p$,

(b) $H(q,p) = \tfrac{1}{2} p^2$,

(c) $H(q,p) = \sin \alpha p$ $\quad (\alpha > 0)$,

for a sufficiently large region of phase space to show the general features of the motion. Where are the fixed points, if any, in each case?

(2) Show that a system whose motion has velocity function

$$(v_x(x, y, t), v_y(x, y, t))$$

is Hamiltonian with coordinate x and conjugate momentum y only if

$$\frac{\partial v_x}{\partial x} = - \frac{\partial v_y}{\partial y} .$$

(3) Describe the qualitative features of the motion of a particle of unit mass moving in the following potentials $V(q)$. In each case sketch the phase diagram and give the equations of the separatrixes, where they exist:

(a) $V(q) = q^{-4} - 2q^{-2}$ $(q > 0)$;

(b) $V(q) = Aq^2 e^{-q^2}$ $\quad (A > 0)$.

(4) Show that the separatrix of the vertical pendulum with Hamiltonian given by equation (4.43) is

$$p = \pm 2\alpha \cos \tfrac{1}{2} \psi .$$

(5) (a) A particle of mass m moves in the arbitrary cubic potential

$$V(q) = Aq^3 + Bq^2 + Cq + D \quad (A > 0).$$

Show that by a suitable linear transformation to a configuration coordinate Q, $V(q)$ can be replaced by a potential

$$W(Q) = AQ^3 + FQ$$

without affecting the motion. Show that F is given by

$$F = C - B^2/(3A).$$

(b) Find any fixed points of the motion in the potential $W(Q)$ for any real value of F and describe the main difference between the motions for $F > 0$ and $F < 0$. What is special about the case $F = 0$?

(6) A particle of mass $m = 1$ kg and position x on a line moves in the potential

$$V(x) = -\tfrac{1}{2} m\alpha^2 x^2,$$

where $\alpha = 1$ (second)$^{-1}$. Obtain an explicit form for the dependence of x on the time t for the motion away from $x = 0$ with $x > 0$ and zero total energy. If the particle starts with this energy at $x_0 = 10^{-9}$ metre, how long to the nearest second will it take to reach $x_1 = 10$ metre?

(7) A particle of unit mass moves in the potential

$$V(q) = -\tfrac{1}{2} q^3.$$

It starts at time $t = 0$ at the point $q = 1$ with momentum $p = 1$. Obtain the subsequent motion; show that it terminates at a time $t = T$ and find T.

(8) A particle of mass m moves in the potential

$$V(q) = -Aq^\gamma \quad (A > 0, \gamma > 0).$$

It starts at time $t = 0$ at the point $q = 1$ with momentum $p = \sqrt{2mA}$. For $\gamma \neq 2$ obtain a relation between q and t for $t > 0$ in terms of A, m and $\beta = 1 - \tfrac{1}{2}\gamma$. For what values of A, m and γ does the motion terminate, and for these values at what time does it terminate? What happens if $\gamma = 2$?

(9) (a) Sketch the phase curve of a particle of mass m moving in the potential

$$V(q) = -Aq^4 \quad (A > 0).$$

Give the equation of the separatrix and find the motion, $q_s(t)$ on the separatrix. Demonstrate that this motion terminates.

(b) The particle starts at time $t = 0$ at the point $q_0 > 0$, with momentum $p_0 > 0$. Obtain an integral expression for the time t at which it reaches the point $q > q_0$, and hence show that the motion terminates.

(10) Sketch the phase diagram, find the fixed points and describe the motion for the Hamiltonian

$$H(q, p) = p + \sqrt{1 - p^2} \sin q \quad (|p| \leqslant 1).$$

(11) Sketch the potential function and contours of the Hamiltonian

$$H(q, p) = \tfrac{1}{2} p^2 + \tfrac{1}{4} q^4 - \tfrac{1}{2} q^2.$$

Give the equation of the separatrix and provide a qualitative description of the motion. Obtain the approximate period of small oscillations in the neighbourhood of any of the stable fixed points.

(12) A particle of mass m has position $r > 0$ and moves under the action of the force

$$F(r) = -kr + \alpha/r^3 \quad (k > 0, \alpha > 0).$$

Determine the potential function $V(r)$ which is zero at the fixed point and provide a qualitative description of the motion. Find the period of the oscillation about the stable fixed point as a function of the energy E for all E at which the motion is oscillatory (see appendix 2).

(13) A system with configuration angle ψ has the Hamiltonian

$$H(\psi,p) = \tfrac{1}{2} p^2 + p \sin \psi \quad (|\psi| \leq \pi).$$

Determine the position and nature of the fixed points. Sketch the phase diagram, showing clearly where there is libration and where there is rotation. Show that the period of the librating motion at energy E is

$$T = 4 \int_0^\alpha \frac{dx}{(\cos^2 x - 2|E|)^{\frac{1}{2}}}$$

where $\cos^2 \alpha = 2|E|$.

(14) An electrically charged particle of mass m is constrained to lie on the straight line between two fixed charges C_1 and C_2 at a distance $2l$ apart. The total force on the particle is the sum of two forces of magnitude μ/r_j^2, where r_j is the distance from the particle to C_j, and each fixed charge repels the particle. Obtain the Hamiltonian and sketch the phase diagram.

The particle starts from rest at a distance kl from the midpoint of C_1 and C_2, where $0 < k < 1$. Show that it librates with period T where

$$T = 2 \left[\frac{ml^3(1-k^2)}{\mu} \right]^{\frac{1}{2}} \int_0^{\frac{1}{2}\pi} d\theta\, (1 - k^2 \sin^2 \theta)^{\frac{1}{2}} \ .$$

(15) Show that the equation of motion

$$\ddot{q} + G(q)\dot{q}^2 - F(q) = 0$$

may be put into Hamiltonian form by choosing the conjugate momentum

$$p = f(q)\dot{q} \quad \text{where} \quad f(q) = \exp\left[2 \int^q dq'\, G(q') \right].$$

Show that the Hamiltonian is

$$H(q,p) = p^2/[2f(q)] + V(q),$$

where

$$V(q) = -\int^q dq'\, F(q') f(q') \ .$$

(16) (a) The motion of a system is defined by the Hamiltonian $H(q, p)$. Show that if a function $f(q, p)$ can be expressed as a differentiable function of H, then the value of $f(q, p)$ is a constant along all phase curves.

(b) If $g(q, p)$ is constant along all phase curves, show that it can be expressed as a function of H.

(17) Consider the rotational motion of a system with configuration angle ψ ($0 \leqslant \psi \leqslant 2\pi$) and Hamiltonian

$H(\psi, p) = p^2/2m + V(\psi)$.

Show that the period T of the rotation can be written as a function of the energy E in the form,

$$T(E) = \sqrt{2m}\ \frac{\mathrm{d}}{\mathrm{d}E} \int_0^{2\pi} \mathrm{d}\psi\, [E - V(\psi)]^{\frac{1}{2}}\ .$$

Hence show that if

$V(\psi) = V_0(\psi) + \epsilon V_1(\psi)$,

then for sufficiently small ϵ the period can be written in the approximate form

$T(E) = T_0(E) + \epsilon T_1(\epsilon)$,

where $T_0(E)$ is the period for $V_0(\psi)$ alone and

$$T_1(E) = -\sqrt{\frac{m}{2}}\ \frac{\mathrm{d}}{\mathrm{d}E} \int_0^{2\pi} \mathrm{d}\psi\ \frac{V_1(\psi)}{[E - V_0(\psi)]^{\frac{1}{2}}}$$

$$= -\frac{\mathrm{d}}{\mathrm{d}E} \int_0^{T_0} \mathrm{d}t\, V_1(\psi_0(t))\ ,$$

where $\psi_0(t)$ is the motion in the potential V_0.

(18) Consider the vibrational motion of a system with configuration coordinate q and Hamiltonian

$H(q, p) = p^2/2m + V(q)$

vibrating between the limits q_1 and q_2 at energy E. Show that the period can be written in the form

$$T(E) = 2\sqrt{2m}\ \frac{\mathrm{d}}{\mathrm{d}E} \int_{q_1}^{q_2} \mathrm{d}q\ [E - V(q)]^{\frac{1}{2}}\ .$$

Hence show that if

$V(q) = V_0(q) + \epsilon V_1(q)$,

then for small ϵ the period can be written in the approximate form

$T(E) = T_0(E) + \epsilon T_1(E)$

with

$$T_1(E) = -\sqrt{2m}\ \frac{\mathrm{d}}{\mathrm{d}E} \int_{q_1^0}^{q_2^0} \mathrm{d}q\ \frac{V_1(q)}{[E - V_0(q)]^{\frac{1}{2}}}$$

where q_1^0 and q_2^0 are the limits of the motion in $V_0(q)$ with energy E. Take particular care with these limits.

(19) Using the results of the previous example, find the approximate change in the period of a linear oscillator with Hamiltonian

$H(q, p) = p^2/2m + \frac{1}{2} m\omega^2 q^2$

due to the small additional potentials,

(a) $\epsilon V_1(q) = \frac{1}{4}\ \epsilon m q^4$,

(b) $\epsilon V_1(q) = \frac{1}{3}\ \epsilon m q^3$.

Only corrections to first order in ϵ are required.

(20) A particle of mass m moves in a one-dimensional potential well $V(q)$ which is unknown, except that it satisfies $V(q) = V(-q)$, $V(0) = 0$. The period of oscillation is measured and its dependence $T(E)$ on the energy E is found. Show that if $q(w)$ is the inverse function of $V(q)$, so that

$w = V(q) \Longleftrightarrow q = g(w)$,

then

$$g(w) = \frac{1}{2\pi}\left(\frac{1}{2m}\right)^{\frac{1}{2}} \int_0^w \mathrm{d}E\ \frac{T(E)}{(w - E)^{\frac{1}{2}}}\ .$$

(21) Consider the system of equations

$\dot{x} = \alpha(y)\,f(x), \quad \dot{y} = \beta(x)\,f(y)$,

where α, β and f are sufficiently well-behaved functions, but are otherwise arbitrary. By defining new variables (q, p),

$x = F(q), y = G(p)$,

show that the system may be put in Hamiltonian form with Hamiltonian,

$$H(q, p) = \int \mathrm{d}G\ \frac{\alpha(G)}{f(G)} - \int \mathrm{d}F\ \frac{\beta(F)}{f(F)}\ ,$$

where $G(p)$ and $F(q)$ are given by the implicit relations

$$q = \int \frac{\mathrm{d}F}{f(F)} \; , \; p = \int \frac{\mathrm{d}G}{f(G)} \; ,$$

(22) A model of neuroelectrical activity gives rise to the equations

$$\dot{x} = (a - by)x\,(1 - x),$$

$$\dot{y} = -(c - dx)y\,(1 - y),$$

where a, b, c and d are positive constants. Using the results of the previous exercise show that this may be put in Hamiltonian form with Hamiltonian

$$H = ap - b \ln (1 + e^p) + cq - d \ln (1 + e^q),$$

where

$$x = \frac{e^q}{1 + e^q} \, , \; y = \frac{e^p}{1 + e^p} \; .$$

Hence show that if $b > a$ and $d > c$ then all the phase curves are closed curves. What happens if either $b < a$ or $c < d$?

5 LAGRANGIANS

5.1 Introduction

In the preceding chapter we found the Hamiltonians of some simple conservative systems directly from Newton's equations of motion by defining the conjugate momenta as

$$p = m\dot{q} = m\,\frac{\mathrm{d}q}{\mathrm{d}t}\,. \tag{5.1}$$

However we cannot always obtain Hamilton's equations in this way, so we need a more general method of obtaining the momenta p.

For example, the equation of motion for a bead sliding smoothly under the influence of gravity on a wire in the shape of the curve $z = \cosh x$, where the z-axis is vertical, is shown in example 5.6 to be

$$\ddot{x}\cosh x + \dot{x}^2 \sinh x + g\tanh x = 0 \tag{5.2a}$$

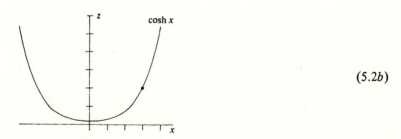

$$\tag{5.2b}$$

Putting $p = m\dot{x}$ gives the pair of equations

$$\left.\begin{aligned} \dot{x} &= v_x = p/m \\ \dot{p} &= v_p = -(m^2 g\tanh x + p^2 \sinh x)/(m\cosh x). \end{aligned}\right\} \tag{5.3}$$

From exercise 4.2 we see that the system is not Hamiltonian in this representation because

$$\frac{\partial v_x}{\partial x} \neq -\frac{\partial v_p}{\partial p}\,. \tag{5.4}$$

It is shown in example 5.6 that equation (5.2) can be put in Hamiltonian form by choosing $p = m \, \dot{x} \sinh^2 x$.

The configuration of a mechanical system of one degree of freedom is represented by a single generalized coordinate q. During the motion q changes and the *generalized velocity* is defined as:

$$\text{generalized velocity} = \frac{dq}{dt} = \dot{q} \,. \tag{5.5}$$

The generalized velocity need not be a physical velocity; for example, if q is an angle, $\dot{q}$ is an angular velocity.

The state of a system can be represented by a point in the two-dimensional phase space in which different coordinate systems may be used. Hamilton's equations describe the motion of a point in the (q, p) coordinate system; Lagrange's equation describes the motion of a point in the $(q, \dot{q})$ coordinate system. In general the transformation between these two coordinate systems, or representations, is not a simple scale change as in equation (5.1) and so the motion in each representation looks quite different; in particular, the motion in the $(q, \dot{q})$ representation is not usually area-preserving. This is illustrated in example 5.6.

In the Hamiltonian formulation q and p are independent variables. Similarly, in the Lagrangian formulation q and $\dot{q}$ are independent variables, so it is convenient to write $u = \dot{q}$, so that (q, u) defines the state of the system.

The configuration of a mechanical system is given as a function of q, and sometimes t. Thus the speed and hence the kinetic energy is naturally expressed as a function of q, $u = \dot{q}$ and t. Because of this it is usually easier to formulate the equations of motion in the (q, u) representation rather than the (q, p) representation. But, since the flow in the (q, p) representation is area-preserving, it is easier to understand.

Thus we need a way of transforming between these two representations. This is provided by the Legendre transformation.

5.2 The Legendre transformation

We can transform between the two representations by obtaining p as a function of q, u and t, $p(q, u, t)$, or finding u as a function of q, p and t, $u(q, p, t)$. In this transformation the variables q and t are fixed so we shall not always write them explicitly. The variable u is defined in terms of p by the first of Hamilton's equations, (4.7a),

$$u = \dot{q} = \frac{\partial H}{\partial p} \,, \tag{5.6}$$

that is, u is the gradient of the Hamiltonian with respect to p.

The graph of the function $H(p)$ may be specified in two ways. First, by explicitly defining the function $H(p)$; second, if $H(p)$ is convex,

$$\frac{\partial^2 H}{\partial p^2} > 0 , \tag{5.7}$$

by defining the function $L(u)$, where $-L(u)$ is the intercept of the tangent having gradient u with the H-axis, as shown in figure 5.1a.

The envelope of the straight lines of gradient u, which cut the H-axis at $-L(u)$, is then the graph of $H(p)$, as in figure 5.1b.

By definition, the relation between $L(u)$ and $H(p)$ is

$$u = [L(u) + H(p)]\,/p. \tag{5.8}$$

We now reintroduce the explicit dependence of all functions on the variables q and t. If $H(q, p, t)$ is the Hamiltonian, the corresponding Lagrangian $L(q, u, t)$ is obtained from equation (5.8):

$$L(q, u, t) = pu - H(q, p, t) , \tag{5.9}$$

where p is given as a function of u, q and t through the implicit relationship

$$u = \frac{\partial H}{\partial p}\,(q, p, t) . \tag{5.10}$$

Fig. 5.1 Figure (a) shows the graph of $H(p)$ with a tangent of gradient $u = \partial H/\partial p$ which intersects the H-axis at $-L(u)$, where L is the Lagrangian. Figure (b) shows how the envelope of these tangents forms the curve $H(p)$.

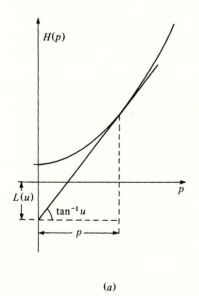

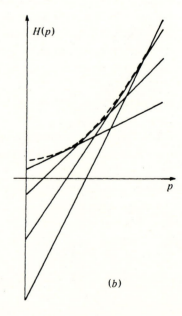

(a)   (b)

Example 5.1

Find the Lagrangians corresponding to the Hamiltonians,

$$(a) \quad H(p) = \frac{1}{2m} p^2 \quad \text{(a free particle)},$$

$$(b) \quad H(p) = \frac{1}{\beta} p^\beta.$$

In the first case, from equation (5.9),

$$L(u) = pu - \frac{p^2}{2m},$$

where p is obtained in terms of u from equation (5.10)

$$u = \frac{\partial H}{\partial p} = \frac{p}{m}, \quad \text{or} \quad p = mu.$$

Thus

$$L(u) = mu^2 - \frac{1}{2} mu^2 = \frac{1}{2} mu^2 = \frac{1}{2} m\dot{q}^2.$$

In the second case

$$L(u) = pu - \frac{p^\beta}{\beta} \quad (u = p^{\beta-1})$$

$$= \frac{1}{\alpha} u^\alpha \quad \text{where} \quad \frac{1}{\alpha} + \frac{1}{\beta} = 1.$$

Example 5.2

Find the Lagrangian corresponding to the Hamiltonian

$$H(q, p, t) = \frac{p^2}{2m} + V(q, t). \tag{5.11}$$

From equations (5.9) and (5.10)

$$L(q, u, t) = pu - \left[\frac{p^2}{2m} + V(q, t) \right] \quad (p = mu),$$

$$= \frac{1}{2} mu^2 - V(q, t)$$

$$= \text{kinetic energy} - \text{potential energy}. \tag{5.12}$$

In order to obtain the Hamiltonian from the Lagrangian we note that the momentum p is the gradient of the Lagrangian with respect to u. This is seen by differentiating equation (5.9) with respect to u, remembering that p is a function of u and that q and t are taken to be constants,

$$\frac{\partial L}{\partial u} = p + \left(u - \frac{\partial H}{\partial p} \right) \frac{\partial p}{\partial u} \tag{5.13}$$

and, since $u = \dot{q} = \partial H / \partial p$,

$$p = \frac{\partial L}{\partial u} (q, u, t). \tag{5.14}$$

As shown in exercise 5.2, $L(u)$ is convex because $H(p)$ is convex. Thus we may construct the Hamiltonian from the Lagrangian in a similar way to the inverse construction: the Hamiltonian $H(p)$ is the negative intercept of the tangent having gradient p with the L-axis.

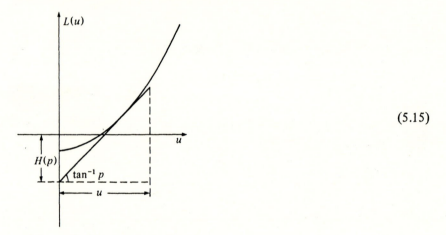

$$\tag{5.15}$$

As before

$$H(q, p, t) = up - L(q, u, t), \tag{5.16}$$

where u is given as a function of q, p and t by solving equation (5.14) as an implicit equation for u.

A comparison of equations (5.16) with (5.9) and (5.14) with (5.10) shows the full symmetry of the relation between the Lagrangian and Hamiltonian.

It is often easier to formulate a problem in mechanics by obtaining a Lagrangian, as shown in section 5.4. Equations (5.14) and (5.16) can then be used to obtain the Hamiltonian.

5.3   The Lagrangian equation of motion

In obtaining the equation of motion in terms of q and u we must remember that the Lagrangian and Hamiltonian are both functions of q and t, which were previously held fixed. Differentiating the Lagrangian, as given by equation (5.9), with respect to q and remembering that p is also a function of q,

$$\frac{\partial L}{\partial q} = u \frac{\partial p}{\partial q} - \frac{\partial H}{\partial q} - \frac{\partial H}{\partial p} \frac{\partial p}{\partial q}$$

$$= -\frac{\partial H}{\partial q} , \qquad (5.17)$$

since $u = \dfrac{\partial H}{\partial p}$.

On using the second of Hamilton's equations (4.7*b*),

$$\frac{dp}{dt} = -\frac{\partial H}{\partial q} = \frac{\partial L}{\partial q} , \qquad (5.18)$$

and since $p = \partial L/\partial u$, from equation (5.14), it follows that

$$\frac{d}{dt} \left(\frac{\partial L}{\partial u} \right) - \frac{\partial L}{\partial q} = 0 , \qquad (5.19)$$

or in terms of $\dot{q}$

$$\frac{d}{dt} \left(\frac{\partial L}{\partial \dot{q}} \right) - \frac{\partial L}{\partial q} = 0 . \qquad (5.20)$$

This is Lagrange's equation of motion. It is a single second-order differential equation, equivalent to Hamilton's pair of coupled first-order equations (4.7). It is usual to use $\dot{q}$ instead of u to represent the generalized velocity. In obtaining partial derivatives it is treated like any other independent variable.

Example 5.3
Find Lagrange's equation of motion for the Lagrangian

$$L(q, \dot{q}, t) = \tfrac{1}{2} m\dot{q}^2 - V(q, t).$$

We have

$$\frac{\partial L}{\partial \dot{q}} = m\dot{q}, \quad \frac{\partial L}{\partial q} = -\frac{\partial V}{\partial q} ,$$

so that (5.20) becomes

$$m\ddot{q} = -\frac{\partial V}{\partial q} ,$$

which is Newton's equation if q is the Cartesian coordinate of a particle. The advantage of the Lagrangian over the Newtonian formulation is that the form of the equation of motion (5.20) is invariant under changes in the generalized coordinate q, as we shall see in the next chapter (but see also exercise 5.7).

Example 5.4

Show that two Lagrangians, $\bar{L}$ and L, differing only by the total time derivative of a function of q and t, describe the same motion.

If

$$\bar{L}(q, \dot{q}, t) = L(q, \dot{q}, t) + \frac{\mathrm{d}}{\mathrm{d}t}\, f(q, t) \qquad (5.21a)$$

$$= L + \dot{q}\,\frac{\partial f}{\partial q} + \frac{\partial f}{\partial t}\,, \qquad (5.21b)$$

then

$$\frac{\partial \bar{L}}{\partial \dot{q}} = \frac{\partial L}{\partial \dot{q}} + \frac{\partial f}{\partial q}\,, \qquad \frac{\partial \bar{L}}{\partial q} = \frac{\partial L}{\partial q} + \frac{\mathrm{d}}{\mathrm{d}t}\left(\frac{\partial f}{\partial q}\right) \qquad (5.22)$$

and hence

$$\frac{\mathrm{d}}{\mathrm{d}t}\left(\frac{\partial \bar{L}}{\partial \dot{q}}\right) - \frac{\partial \bar{L}}{\partial q} = \frac{\mathrm{d}}{\mathrm{d}t}\left(\frac{\partial L}{\partial \dot{q}}\right) - \frac{\partial L}{\partial q}\,. \qquad (5.23)$$

Thus if $q(t)$ satisfies the equations of motion obtained from $\bar{L}$ it also satisfies those obtained from L.

This transformation is often useful in simplifying problems with time-dependent Lagrangians; see for instance example 5.7.

Note that the two Lagrangians define different momenta, and hence different Hamiltonians. If $p = \partial L/\partial \dot{q}$ and $\bar{p} = \partial \bar{L}/\partial \dot{q}$ then from (5.21b)

$$\bar{p} = p + \frac{\partial f}{\partial q}\,. \qquad (5.24)$$

The momenta p and $\bar{p}$ are the same only if f is independent of q.

The relationship between the Hamiltonians $\bar{H}(q, \bar{p}, t)$ and $H(q, p, t)$ generated from $\bar{L}$ and L is given in the next chapter, but see example 5.7 and exercises 5.18 and 5.19 for specific applications.

5.4 Formulation

For those dynamical systems having a kinetic energy $T(q, \dot{q}, t)$ and a potential energy $V(q, t)$ the Lagrangian is given by

$$L(q, \dot{q}, t) = \text{kinetic energy} - \text{potential energy}$$

$$= T(q, \dot{q}, t) - V(q, t) \qquad (5.25)$$

and the Hamiltonian is obtained directly from (5.16).

In many cases the kinetic energy is quadratic in the generalized velocity $\dot{q}$,

$$T = \tfrac{1}{2}\, m\, G(q, t)\, \dot{q}^2, \qquad (5.26)$$

G being some non-zero function of q and t only. In this case the momentum is

$$p = \frac{\partial L}{\partial \dot{q}} = \frac{\partial T}{\partial \dot{q}} = mG(q,t)\dot{q} \tag{5.27}$$

and from (5.16) the Hamiltonian is

$$H(q,p,t) = \frac{p^2}{2mG(q,t)} + V(q,t)$$

$$= \text{kinetic energy} + \text{potential energy}, \tag{5.28}$$

so the kinetic energy is then quadratic in the momentum also.

If neither $G(q,t) = G(q)$ nor $V(q,t) = V(q)$ depend explicitly upon time the Hamiltonian is the total energy of the system and is conserved. Then the contour methods described in chapter 4 may be used to obtain a qualitative under-standing of the motion. If one or both of G and V depend explicitly upon time, the methods described in chapters 9 and 10 may be applicable.

For most of the examples we consider, the kinetic energy of a particle of mass m is given by $\frac{1}{2}mw^2$ where w is its speed. However the particle is often constrained by a rod or wire, so the relation between w and the configuration coordinate q or ψ is not always simple.

The gravitational potential energy of a particle in the neighbourhood of the surface of the Earth is usually taken as mgh, where h is the height and g the acceleration due to gravity, taken as constant. There may be many sources of potential energy, like two centres of force. In that case the total potential energy is always taken as the sum of the values that each would contribute on its own.

Example 5.5
Find the Lagrangian and Hamiltonian for a pendulum moving in a vertical plane.

Here the configuration of the system may be defined by the angle between the pendulum and the downward vertical, ψ. The generalized velocity $u = \dot{\psi}$ is the angular velocity of OA.

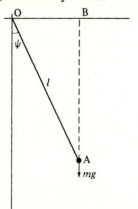

The kinetic energy of the mass m at A is

$$T = \tfrac{1}{2}m \text{ (speed)}^2 = \tfrac{1}{2}mw^2 = \tfrac{1}{2}ml^2u^2$$

and the potential energy is

$$V = -mg\,AB = -mgl\cos\psi,$$

so that the Lagrangian is

$$L(\psi, u) = \tfrac{1}{2}ml^2u^2 + mgl\cos\psi. \tag{5.29}$$

The momentum conjugate to ψ is

$$p = \frac{\partial L}{\partial u} = ml^2u \quad \text{or} \quad u = \dot\psi = \frac{1}{ml^2}\,p.$$

In this case p is the angular momentum of the mass m about an axis through O and perpendicular to the page.

The Hamiltonian, from Equation (5.16), is

$$H(\psi, p) = pu - (\tfrac{1}{2}ml^2u^2 + mgl\cos\psi) \qquad \left(u = \frac{p}{ml^2} \right),$$

$$= \frac{p^2}{2ml^2} - mgl\cos\psi. \tag{5.30}$$

The motion determined by this Hamiltonian is described in section 4.7.

Example 5.6
Find the Lagrangian and Hamiltonian for the motion of a bead of mass m sliding smoothly on a wire in the shape $z = f(x)$, the z-axis and x-axis being respectively vertical and horizontal. Consider in detail the particular case $f(x) = \cosh x$.

In this example x may be used for the generalized coordinate and the generalized velocity is $\dot x$, the velocity parallel to the x-axis.

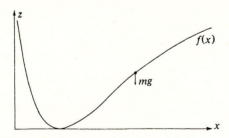

The potential energy is

$$V(x) = mgz = mgf(x)$$

and the kinetic energy is

$$T = \tfrac{1}{2} m \text{ (speed)}^2 = \tfrac{1}{2} m \left(\dot{x}^2 + \dot{z}^2 \right).$$

but $\dot{z} = \dot{x} \, df/dx$ so that

$$T(x, \dot{x}) = \tfrac{1}{2} m \dot{x}^2 \left[1 + f'(x)^2 \right],$$

which is of the form given in equation (5.26).

The Lagrangian is

$$L = \tfrac{1}{2} m \dot{x}^2 \left[1 + f'(x)^2 \right] - mgf(x), \tag{5.31}$$

and the momentum conjugate to x is

$$p = \frac{\partial L}{\partial \dot{x}} = m \dot{x} \left[1 + f'(x)^2 \right],$$

which is a function of both $\dot{x}$ and x.

Fig. 5.2 Diagram showing the contours of the Hamiltonian (5.33) in the (x, p)- and $(x, \dot{x})$-representations. The flow in the (x, p)-representation is area-preserving but it is not in the $(x, \dot{x})$-representation.

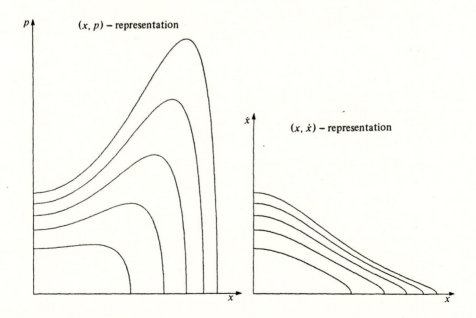

The Hamiltonian is then from (5.16)

$$H(x,p) = \dot{x}p - \left\{ \tfrac{1}{2} m \dot{x}^2 \left[1 + f'(x)^2 \right] - mgf(x) \right\} \quad \left(\dot{x} = \frac{p}{m[1 + f'(x)^2]} \right),$$

$$= \frac{p^2}{2m \left[1 + f'(x)^2 \right]} + mgf(x). \tag{5.32}$$

Since $f(x)$ is time-independent, H is a constant of the motion. In the particular case $z = \cosh x$, we have

$$H(x,p) = \frac{p^2}{2m \cosh^2 x} + mg \cosh x. \tag{5.33}$$

This defines the (x, p)-contours, as shown in figure 5.2, with x and p as Cartesian coordinates. Also shown are the $(x, \dot{x})$-contours which satisfy the equation

$$E = \tfrac{1}{2} m \dot{x}^2 \cosh^2 x + mg \cosh x, \tag{5.34}$$

obtained by expressing the energy in terms of $(x, \dot{x})$. It is seen that the motion looks quite different in each representation.

Example 5.7
Find the Lagrangian and Hamiltonian of a pendulum comprising a mass m attached to a light, stiff rod AB of length l free to move in a vertical plane, and such that the end A of the rod is forced to move vertically, its distance from a fixed point O being a given function $\gamma(t)$ of time.

The generalized coordinate is the angle ψ between AB and the downward vertical. In order to determine the speed of the mass, it is helpful to write down its coordinates with respect to the axes Ox and Oz shown in the diagram.

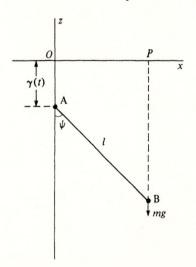

These are

$$x = l \sin \psi,$$

$$z = -l \cos \psi - \gamma(t).$$

The potential energy is

$$V(z, t) = mgz = -mg \left[l \cos \psi + \gamma(t) \right]$$

and the kinetic energy is

$$T = \tfrac{1}{2} m \text{ (speed)}^2 = \tfrac{1}{2} m (\dot{x} + \dot{z})^2$$

$$= \tfrac{1}{2} m \left[l^2 \dot{\psi}^2 \cos^2 \psi + (l \dot{\psi} \sin \psi - \dot{\gamma})^2 \right]$$

$$= \tfrac{1}{2} m \left[l^2 \dot{\psi}^2 - 2l \dot{\psi} \dot{\gamma} \sin \psi + \dot{\gamma}^2 \right].$$

Thus the Lagrangian is

$$L = \tfrac{1}{2} m (l^2 \dot{\psi}^2 - 2l \dot{\psi} \dot{\gamma} \sin \psi) + mgl \cos \psi + h(t), \qquad (5.35)$$

where

$$h(t) = \tfrac{1}{2} m \dot{\gamma}^2 + mg \gamma$$

is a function of t only and, as shown in example 5.4, may be ignored.

The momentum conjugate to ψ is

$$p = \frac{\partial L}{\partial \dot{\psi}} = m (l^2 \dot{\psi} - l \dot{\gamma} \sin \psi)$$

and, after some manipulation, we find that the Hamiltonian is

$$H(\psi, p, t) = \frac{(p + ml \dot{\gamma} \sin \psi)^2}{2ml^2} - mgl \cos \psi. \qquad (5.36)$$

An alternative, but more convenient, Hamiltonian may be obtained by using the result of example 5.4. On writing

$$\dot{\psi} \dot{\gamma} \sin \psi = -\frac{\mathrm{d}}{\mathrm{d}t} (\dot{\gamma} \cos \psi) + \ddot{\gamma} \cos \psi$$

we obtain

$$\bar{L} = \tfrac{1}{2} ml^2 \dot{\psi}^2 + ml(g - \ddot{\gamma}) \cos \psi, \qquad (5.37)$$

showing that the vertical acceleration has the same effect as a time-varying ·gravitational field. The conjugate momentum is now

$$\bar{p} = \frac{\partial \bar{L}}{\partial \dot{\psi}} = ml^2 \dot{\psi}$$

and the Hamiltonian,

$$\bar{H} = \frac{\bar{p}^2}{2ml^2} - ml(g - \ddot{\gamma}) \cos \psi. \qquad (5.38)$$

The general theory of the following chapter will show how H and $\bar{H}$ are related. The theory of chapter 9 will show how in some cases approximate solutions to this time-dependent problem may be found.

Exercises for chapter 5

(1) Use the Legendre transformation to obtain the Hamiltonian functions $H(p)$ corresponding to the Lagrangians

    (a) $L(u) = e^u$,

    (b) $L(u) = \tan u$ $(0 \leqslant u < \frac{1}{2}\pi)$.

(2) If $H(p)$ and $L(u)$ are the Hamiltonian and Lagrangian for a system, show that

$$\frac{\partial^2 H}{\partial p^2}\frac{\partial^2 L}{\partial u^2} = 1.$$

What can be deduced about the convexity of $H(p)$ and $L(u)$?

(3) The Lagrangian

    $L(u) = \sin u$ $(0 \leqslant u < \frac{1}{2}\pi)$

is not convex. Define a modified Legendre transformation for such a Lagrangian and apply it to find $H(p)$ in this case.

(4) Show that

$$\frac{\partial H}{\partial t}(q,p,t) = -\frac{\partial L}{\partial t}(q,\dot{q},t)$$

and deduce that, if the Lagrangian does not depend explicitly on time, then the value of the Hamiltonian is constant on all phase curves. Is the Lagrangian constant on all phase curves?

(5) Find the Hamiltonians corresponding to the Lagrangians

    (a) $L(\dot{q}) = (1 - \dot{q}^2)^{\frac{1}{2}}$ $(|\dot{q}| \leqslant 1)$,

    (b) $L(q,\dot{q},t) = \frac{1}{2}e^{\alpha t}(\dot{q}^2 - \omega^2 q^2)$.

Find the Lagrangian corresponding to the Hamiltonian

    (c) $H(q,p) = \frac{1}{2}p^2 + p \sin q$.

(6) Sketch the phase curves in the $(q,\dot{q})$ and (q,p) representations for the system whose Lagrangian is

$$L = q\dot{q} - (1 - \dot{q}^2)^{\frac{1}{2}} \quad (|\dot{q}| \leqslant 1).$$

(7) Two configuration coordinates Q and q of a mechanical system are related by the equation

$$q = f(Q).$$

If $L(q, \dot{q}, t)$ is the Lagrangian in q-representation, show by direct evaluation that Lagrange's equation in the Q-representation is

$$\frac{d}{dt} \left(\frac{\partial \bar{L}}{\partial \dot{Q}} \right) - \left(\frac{\partial \bar{L}}{\partial Q} \right) = 0.$$

where

$$\bar{L}(Q, \dot{Q}, t) = L(f(Q), f'(Q)\dot{Q}, t).$$

Furthermore show that the momenta p and P conjugate to q and Q are related by

$$p = P/f'(Q),$$

(8) If a particle of mass m is projected vertically upwards, far from the surface of the Earth, the Lagrangian is

$$L = \tfrac{1}{2} m \dot{z}^2 + GMm/(R + z),$$

G being the gravitational constant, M and R the mass and radius of the Earth and z the height of the particle above the surface. Find the Hamiltonian and sketch the phase diagram showing clearly the region where the motion is bounded and the region where it is unbounded. Find the equation of the separatrix. For the bounded motion find the time taken to return to the surface.

(9) A particle of mass m is constrained to move in a vertical plane under the influence of gravity along a given curve with parametric equations

$$z = h(\eta), \quad x = f(\eta)$$

where the z-axis is vertically upwards. Taking η as the generalized coordinate show that the Lagrangian is

$$L = \tfrac{1}{2} m \dot{\eta}^2 \left[\left(\frac{dh}{d\eta} \right)^2 + \left(\frac{df}{d\eta} \right)^2 \right] - mgh(\eta),$$

and find the corresponding Hamiltonian.

(10) A particle of mass m is constrained to move in a vertical plane under the influence of gravity on the smooth cycloid with parametric equation

$$z = A (1 - \cos \phi), \quad x = A (\phi + \sin \phi) \quad (-\pi < \phi < \pi),$$

where A is a positive constant and the z-axis is vertically upwards. If s

is the arc length along the cycloid, measured from the bottom of the curve, show that the Lagrangian of the motion is

$$L = \tfrac{1}{2} m\dot{s}^2 - mgs^2/8A.$$

Hence show that the period of the motion is independent of the amplitude. Find this period.

(11) Obtain the Lagrangian and Hamiltonian for the motion of a particle of mass m swinging under gravity when attached to a fixed point O by a light straight string of given variable length $l(t)$, where its motion is confined to a vertical plane through O.

(12) Find the Hamiltonian and Hamilton's equations for the motion of a particle of mass m sliding under gravity on a smooth rigid parabolic wire in the vertical x - z plane where the z-axis is vertically upwards. Consider the cases:

(*a*) the parabolic wire is fixed so that it lies on the curve

$$z = \tfrac{1}{2} \alpha^2 x^2 ;$$

(*b*) The wire moves horizontally, so that its equation is

$$z = \tfrac{1}{2} \alpha^2 [x - \gamma(t)]^2 ,$$

where $\gamma(t)$ is a given function of time.

(13) A particle of mass m moves under the influence of gravity along the smooth spiral wire defined, through the parameter ψ, by

$$\left. \begin{aligned} z &= k\psi, \\ x &= a \cos \psi, \\ y &= a \sin \psi, \end{aligned} \right\} \quad (a > 0, k > 0),$$

where a and k are constant and the z-axis is vertically upwards. Obtain Hamilton's equations of motion and find $\psi(t)$ in terms of $\psi(0)$ and $\dot{\psi}(0)$. What are the parametric equations for the spiral if, instead of being fixed, it rotates about the z-axis with constant angular velocity Ω? Obtain the expression for the speed w in terms of ψ and $\dot{\psi}$ and hence find the equations of motion and their solution $\psi(t)$ for this new situation.

(14) A bead of mass m is constrained to slide under the influence of gravity on a smooth parabolic wire with equation

$$z = \tfrac{1}{2} \alpha^2 x^2 ,$$

where the z-axis is vertically upwards. The wire rotates about the z-axis with constant angular velocity Ω. Show that the Hamiltonian is

$$H = \frac{p^2}{2m(1 + \alpha^4 x^2)} + \frac{mx^2}{2} (g\alpha^2 - \Omega^2).$$

Sketch the phase curves and give a qualitative description of the
motion, considering the cases $g\alpha^2 > \Omega^2$ and $g\alpha^2 < \Omega^2$ separately. In
particular, show that if $g\alpha^2 > \Omega^2$ the bead always librates with period

$$T = \frac{2\sqrt{2m}}{\beta} \int_0^{\frac{1}{2}\pi} d\theta \left(1 + \frac{\alpha^4 E}{\beta^2} \sin^2 \theta \right)^{\frac{1}{2}},$$

where E is the energy and $\beta^2 = \frac{1}{2} m(g\alpha^2 - \Omega^2)$.

(15) A particle is constrained to slide under gravity on a smooth wire in the
shape of a vertical circle with radius R, when the wire rotates about a
vertical diameter with constant angular velocity Ω. Show that the
Hamiltonian is

$$H = \frac{p^2}{2mR^2} - m \left(\frac{1}{2} R^2 \Omega^2 \sin^2 \psi + gR \cos \psi \right),$$

where ψ is the angular displacement of the particle from the downward vertical, with the centre of the circle as origin. Determine the
fixed points of the system, discuss their stability and sketch the phase
diagram. Determine the frequency of small oscillations about any
elliptic fixed points.

(16) A particle of mass m is embedded, at a distance l from the axis, in a
massless cylinder of radius R. The cylinder rolls without slipping on a
plane inclined at an angle α to the horizontal. The axis of the cylinder
remains horizontal. The particle is influenced by gravity. Find the
Hamiltonian and show that, if $R \sin \alpha < l$, the system has a stable fixed
point. Find the period of small oscillations about this fixed point.

(17) A pendulum is constructed by attaching a mass m to an inextensible
string of length l. The other end of the string is attached at A to the top
of a horizontal cylinder of radius R, $(R < 2l/\pi)$, as shown in the diagram.

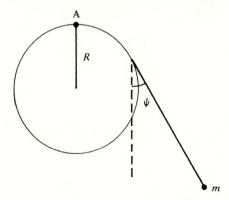

Assuming that the motion is confined to a vertical plane through A perpendicular to the cylinder axis, and that the string makes an angle ψ with the downward vertical, show that the Hamiltonian is

$$H(\phi, p) = \frac{p^2}{2m(R\phi - l)^2} - mg\left[R(1 - \cos \phi) + (l - R\phi) \sin \phi\right],$$

where $\phi + \psi = \frac{1}{2}\pi$ and

$$p = m\,\dot\phi\,(R\phi - l).$$

Hence show that the angular frequency of small oscillations about the stable fixed point is

$$\omega = \left[\,g/(l - \tfrac{1}{2}R\pi)\,\right]^{\frac{1}{2}}.$$

(18) Suppose a plane vertical pendulum is made of a mass m attached to one end of a light rod of length l, where the other end moves horizontally in the plane of the motion, so that its distance from a fixed point is given by a known function, $\gamma(t)$, of time t. Show that the Hamiltonian can be written

$$H(\psi, p) = p^2/2ml^2 + ml[\beta(t) \sin \psi - g \cos \psi],$$

where $\beta(t) = \ddot\gamma(t)$ and ψ is the angle between the pendulum and the downward vertical. For the particular case in which $\beta = ag$, where a is a constant, determine the fixed points, discuss their stability, and find the period of small oscillations about any elliptic fixed points.

(19) (a) If a system has the Lagrangian

$$L = \tfrac{1}{2}\,G(q, t)\dot{q}^2 + F(q, t)\dot{q} - V(q, t),$$

show that the Hamiltonian has the form

$$H = \frac{[p - F(q, t)]^2}{2G(q, t)} + V(q, t),$$

where

$$p = G\dot{q} + F.$$

(b) Show that the Lagrangian

$$\bar{L} = \tfrac{1}{2}\,G(q, t)\dot{q}^2 - [V(q, t) + \partial f(q, t)/\partial t] \quad (\partial f/\partial q = F)$$

gives the same equations of motion as the Lagrangian L. Show that the Hamiltonian $\bar{H}$ corresponding to $\bar{L}$ is

$$\bar{H} = \frac{\bar{p}^2}{2G(q, t)} + V(q, t) + \frac{\partial f}{\partial t}\,(q, t),$$

where

$$\bar{p} = G\dot{q}$$

(c) Show explicitly that H and $\bar{H}$ give rise to the same motion.

(20) The Lagrangian for a free relativistic particle moving along a straight line with speeds close to the velocity of light, c, is given by

$$L(\dot{q}) = -mc^2 [1 - (\dot{q}/c)^2]^{\frac{1}{2}},$$

where q is the position on the line. Suppose a relativistic particle is acted upon by a force which contributes a potential energy

$$V(q) = \tfrac{1}{2} m\omega^2 q^2.$$

Obtain the momentum p and the Hamiltonian $H(q, p)$. Is the value $E = H(q, p)$ conserved along a phase curve? What is its value at zero velocity and at the origin? Show that the period of oscillation may be expressed in the form

$$T = \frac{4}{c\omega} \sqrt{\frac{2}{m}} \int_0^{\frac{1}{2}\pi} d\theta \, \frac{E - (E - mc^2) \sin^2 \theta}{[(E + mc^2) - (E - mc^2) \sin^2 \theta]^{\frac{1}{2}}}.$$

Confirm that this period approaches its non-relativistic value when the maximum value of $|\dot{q}|$ along a phase curve is small compared to c.

6 TRANSFORMATION THEORY

6.1 Introduction

This chapter deals with Hamiltonian theory. Even within this theory there are many representations with different generalized coordinates and conjugate momenta.

A good choice of coordinate system, or representation, leads to simplification. If we start with a bad initial (q, p) we want to transform to a good final (Q, P). It is helpful if the transformation preserves the Hamiltonian form of the equations of motion, and particularly helpful if the Hamiltonian in the (Q, P) representation is a function of P only.

Thus the two main aims of transformation theory are:

(T1) to determine those transformations which preserve the form of Hamilton's equations; and

(T2) to find those particular transformations which reduce the Hamiltonian to a simple form.

The main object of this chapter is to solve T1 and to establish a formalism, used in later chapters, useful for the solution of T2.

We first consider time-independent transformations, for which the theory is slightly easier. The theory is extended in section 6.6 to include time-dependent transformations.

6.2 The theory of time-independent transformations

In chapter 4 we saw that the flow obtained from Hamilton's equations is area-preserving. We wish to perform transformations between different representations, or coordinate systems in phase space, so that for the first time we need a definition of area for more than one representation; the form of this definition must be independent of the representation.

If a region R in the (q, p) representation is transformed into a region S in the (Q, P) representation, the areas in each representation are *defined*, with (q, p) and (Q, P) as Cartesian coordinates, thus:

$$A_R = \iint_R dq\, dp \; ; \qquad A_S = \iint_S dQ\, dP . \tag{6.1}$$

$$(6.2)$$

For general transformations A_R and A_S can have different numerical values since

$$\iint_S dQ\,dP = \iint_R dq\,dp\ \frac{\partial(Q,P)}{\partial(q,p)},\tag{6.3}$$

where the second integrand is a Jacobian determinant.

For the flow in each representation to be Hamiltonian, i.e. area-preserving, it is necessary that the numerical value of all areas be representation-independent. That is, we require the Jacobian determinant to be constant. As shown in exercise 6.14, the value of this constant may be chosen as unity without loss of generality,

$$\frac{\partial(Q,P)}{\partial(q,p)} = \frac{\partial Q}{\partial q}\frac{\partial P}{\partial p} - \frac{\partial Q}{\partial p}\frac{\partial P}{\partial q} = 1,\tag{6.4a}$$

or equivalently

$$\frac{\partial(q,p)}{\partial(Q,P)} = \frac{\partial q}{\partial Q}\frac{\partial p}{\partial P} - \frac{\partial q}{\partial P}\frac{\partial p}{\partial Q} = 1.\tag{6.4b}$$

We name a transformation satisfying this condition *canonical*. For example the transformation

$$Q = p,\quad P = -q\tag{6.5}$$

has a Jacobian 1 and so is canonical. The transformation from polar coordinates to Cartesian coordinates has the form

$$q = P\cos Q,\quad p = P\sin Q.\tag{6.6}$$

The Jacobian is not constant so the transformation is not canonical.

Example 6.1

Determine the values of α and β for which the following is a canonical transformation.

$$Q = q + \alpha p$$
$$P = p + \beta q.\tag{6.7}$$

The Jacobian of the transformation is

$$\frac{\partial(Q,P)}{\partial(q,p)} = \begin{vmatrix} 1 & \alpha \\ \beta & 1 \end{vmatrix} = 1 - \alpha\beta,$$

which equals unity when either α or β is zero. For $\beta = 0$

$$
\begin{aligned}
Q &= q + \alpha p \\
P &= p,
\end{aligned}
\tag{6.8}
$$

which is a shear in the q-direction, as shown in (6.9).

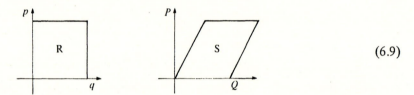

(6.9)

When $\alpha = 0$ we obtain a shear in the p-direction, and when $\alpha = \beta = 0$ we obtain the identity transformation.

The transformation $(q,p) \rightarrow (Q,P)$ may be interpreted in two distinct ways. In one case the phase point and the state that it represents change but the co-ordinate system is unchanged; the phase point moves from (q,p) to (Q,P) in the same coordinate system. This interpretation is shown in figure 6.1a, where a region R is transformed into a region S of different shape. Alternatively the coordinate system changes but the phase point and the state that it represents are unchanged, as shown in figure 6.1b.

Fig. 6.1 Sketch illustrating the different interpretation of the $(q,p) \rightarrow (Q,P)$ transformation. In (a) the phase point A at (q,p) moves to B at (Q,P), so that the circular region R moves to the oval region S. In (b) the phase point A is unchanged but the rectangular coordinate system (q,p) is changed into the curvilinear coordinate system (Q,P)

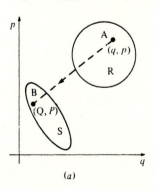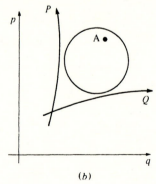

(a)   (b)

Consider for example the transformation of equation (6.8). The first interpretation in which the phase points change is shown in (6.9) where the square is transformed into the parallelogram. The second interpretation gives a skew coordinate system where the Q-axis, the line $P = 0$, coincides with the q-axis and the P-axis, the line $Q = 0$, is the line $\alpha p = -q$ as shown in (6.10).

$$(6.10)$$

Although these interpretations are quite different, the mathematics is the same; the different interpretations apply to different circumstances.

6.3 The $F_1(Q, q)$ generating function

A general transformation between two representations requires two functions for its specification, but, because of the area-preserving property, only one function is necessary to specify a canonical transformation.

The area enclosed by a curve $\mathscr{C}$ in phase space may be expressed either as a double or, using Stokes' theorem, a single integral;

$$\iint_R dq\,dp = \oint_{\mathscr{C}} p(q)\,dq, \qquad (6.11)$$

where the curve $\mathscr{C}$ is traversed clockwise.

It is now helpful to consider the transformation $(q, p) \to (Q, P)$ as defining a new coordinate system on the phase space, as discussed at the end of the preceding section. The area may be expressed as a line integral in each coordinate system and, since the Jacobian of the transformation is unity,

$$\oint_{\mathscr{C}} p\,dq = \oint_{\mathscr{C}} P\,dQ, \qquad (6.12)$$

$\mathscr{C}$ being an arbitrary closed curve in the phase space. But we may choose almost *any* pair of variables to be independent; in particular we may take (Q, q) as independent variables and express P and p as functions of them. Then equation (6.12) may be written as the line integral

$$\oint_{\mathscr{C}} [p(Q, q)\,dq - P(Q, q)\,dQ] = 0. \qquad (6.13)$$

Since this equation is true for all curves $\mathscr{C}$ the integrand must be a perfect differential of a function $F_1(Q, q)$, that is

$$\oint_{\mathscr{C}} [p\,dq - P\,dQ] = \oint_{\mathscr{C}} dF_1(Q, q)$$

$$= \oint_{\mathscr{C}} \frac{\partial F_1}{\partial q}\,dq + \oint_{\mathscr{C}} \frac{\partial F_1}{\partial Q}\,dQ. \tag{6.14}$$

Because the curve $\mathscr{C}$ is arbitrary the coefficients of dq and dQ may be equated to give

$$p = \frac{\partial F_1}{\partial q}(Q, q), \quad P = -\frac{\partial F_1}{\partial Q}(Q, q). \tag{6.15}$$

That is the area-preserving property of the canonical transformation implies the existence of $F_1(Q, q)$, which is named a *generating function*. It is a particularly compact way of representing the transformation $(q, p) \rightarrow (Q, P)$. The generating function is undefined to within an additive constant.

The transformation $(q, p) \rightarrow (Q, P)$ is given implicitly by (6.15) and to obtain it explicitly an inversion is necessary. This is always possible if

$$\frac{\partial^2 F_1}{\partial q \partial Q}(Q, q) \neq 0, \tag{6.16}$$

which is a necessary and sufficient condition that $F_1(Q, q)$ should generate a canonical transformation. In practice this inversion can be awkward, and frequently $F_1(Q, q)$ is multivalued, giving more than one solution.

Coordinate transformations in configuration space, for example $Q = e^q$, are often valuable, but in this case the previous form of the generating function cannot be used because q and Q are not independent and so it is not possible to express either p or P in terms of them. In this case the $F_1(Q, q)$ generating function does not exist, but using (6.12) we have,

$$\oint_{\mathscr{C}} dq\,\left(p - P\frac{dQ}{dq}\right) = 0 \tag{6.17}$$

or, since the curve $\mathscr{C}$ is arbitrary,

$$p = P\frac{dQ}{dq}. \tag{6.18}$$

Such transformations are important; for example the identity transformation

$$Q = q, \quad P = p \tag{6.19}$$

is of this type. In the next section we see how to find generating functions for this type of transformation.

Example 6.2

Show explicitly that a transformation generated by $F_1(Q, q)$ is area-preserving.

Since $p = \partial F_1 / \partial q$ and $P = -\partial F_1 / \partial Q$ it is necessary to express the Jacobian (6.4) in terms of the independent variables (q, Q). To do this we use standard properties of the Jacobian:

$$\frac{\partial(Q, P)}{\partial(q, p)} = \frac{\partial(Q, P)}{\partial(q, Q)} \frac{\partial(q, Q)}{\partial(q, p)} = \frac{\partial(Q, P)}{\partial(q, Q)} \left[\frac{\partial(q, p)}{\partial(q, Q)} \right]^{-1}$$

$$= -\frac{\partial P}{\partial q}\bigg|_Q \left[\frac{\partial p}{\partial Q}\bigg|_q \right]^{-1}$$

$$= \frac{\partial^2 F_1}{\partial q \partial Q} \left[\frac{\partial^2 F_1}{\partial Q \partial q} \right]^{-1} = 1.$$

Example 6.3

The transformation $(q_1, p_1) \rightarrow (q_2, p_2)$, where $q_i = q(t_i)$, $p_i = p(t_i)$ $(i = 1, 2)$ are points on a phase curve, is area-preserving and therefore may be represented by a generating function $F_1(q_2, q_1)$ with

$$p_1 = \frac{\partial F_1}{\partial q_1}(q_2, q_1), \quad p_2 = -\frac{\partial F_1}{\partial q_2}(q_2, q_1).$$

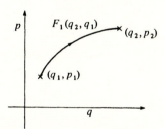

Note that generating functions can produce transformations which are either changes of state or changes of coordinate system. In this case we have a change of state.

6.4   Other forms of generating function

The generating function $F_1(Q, q)$ depended upon Q and q, rather than another pair of variables, because we expressed equation (6.12) in terms of (Q, q) instead of (P, q), (Q, p) or (P, p). For any closed curve $\mathscr{C}$

$$0 = \oint_{\mathscr{C}} \mathrm{d}(PQ) = \oint_{\mathscr{C}} P \, \mathrm{d}Q + \oint_{\mathscr{C}} Q \, \mathrm{d}P, \tag{6.20}$$

so that

$$\oint_{\mathscr{C}} P \, dQ = - \oint_{\mathscr{C}} Q \, dP, \tag{6.21}$$

with similar relations for (q, p). Hence each side of equation (6.12) may be expressed in two ways, and there are four different types of generating function, depending upon the choice of independent variable: $F_1(Q, q)$, $F_2(P, q)$, $F_3(Q, p)$ and $F_4(P, p)$. We present the theory for $F_2(P, q)$: the results for all generating functions are summarized in table 6.1.

Using (P, q) as independent variables, the equations (6.21) and (6.12) give, in place of equation (6.14)

$$0 = \oint_{\mathscr{C}} (p \, dq + Q \, dP). \tag{6.22}$$

Then, since $\mathscr{C}$ is arbitrary,

$$\oint_{\mathscr{C}} (p \, dq + Q \, dP) = \oint_{\mathscr{C}} dF_2(P, q) \tag{6.23}$$

and

$$p = \frac{\partial F_2}{\partial q}(P, q), \quad Q = \frac{\partial F_2}{\partial P}(P, q). \tag{6.24}$$

Table 6.1. *Table showing how the dependent variables are related to the various generating functions*

Generating function	Dependent variables			
	q	p	Q	P
$F_1(Q, q)$		$\dfrac{\partial F_1}{\partial q}$		$-\dfrac{\partial F_1}{\partial Q}$
$F_2(P, q)$		$\dfrac{\partial F_2}{\partial q}$	$\dfrac{\partial F_2}{\partial P}$	
$F_3(Q, p)$	$-\dfrac{\partial F_3}{\partial p}$			$-\dfrac{\partial F_3}{\partial Q}$
$F_4(P, p)$	$-\dfrac{\partial F_4}{\partial p}$		$\dfrac{\partial F_4}{\partial P}$	

Note that, for this generating function, both equations have the same sign. Again $F_2(P, q)$ is undefined to within an additive constant.

The $F_1(Q, q)$ and $F_2(P, q)$ generating functions may be related using the identity (6.20) to rewrite equation (6.23) as

$$\oint_{\mathscr{C}} dF_2 = \oint_{\mathscr{C}} p\, dq - P\, dQ + d(PQ) = \oint_{\mathscr{C}} d(F_1 + PQ). \qquad (6.25)$$

Hence the same canonical transformation is generated by $F_1(Q, q)$ or by $F_2(P, q)$, where

$$F_2(P, q) = F_1(Q, q) + PQ \qquad (6.26)$$

and where Q on the right-hand side is expressed as a function of (P, q) using the implicit equation

$$P = -\frac{\partial F_1}{\partial Q}(Q, q). \qquad (6.27)$$

Example 6.4
The identity transformation is generated by

$$F_2(P, q) = Pq. \qquad (6.28)$$

Example 6.5
Show that the transformation

$$Q = \ln\left(\frac{\sin p}{q}\right), \quad P = q \cot p \qquad (6.29)$$

is canonical and determine the generating functions $F_1(Q, q)$ and $F_2(P, q)$.

We have

$$P\, dQ = q \cot p \left[(\cot p)\, dp - \frac{dq}{q} \right] \qquad (6.30)$$

and consequently

$$p\, dq - P\, dQ = d(pq + q \cot p), \qquad (6.31)$$

showing that the integrand of equation (6.13) is a perfect differential. The $F_1(Q, q)$ generating function is found by expressing $pq + q \cot p$ in terms of (Q, q) using the relation

$$p = \sin^{-1}(q\, e^Q)$$

obtained from equation (6.29). Thus

$$p\, dq - P\, dQ = dF_1(Q, q), \qquad (6.32)$$

where

$$F_1(Q, q) = q \sin^{-1}(q \ e^Q) + [e^{-2Q} - q^2]^{1/2}. \tag{6.33}$$

The existence of $F_1(Q, q)$ shows that the transformation is canonical as shown in example 6.2. Using the relation between F_1 and F_2, equation (6.26),

$$F_2(P, q) = qp + q \cot p + PQ, \tag{6.34}$$

where the right-hand side is to be expressed in terms of P and q only. We have, from equation (6.29),

$$p = \tan^{-1}(q/P) \tag{6.35a}$$

$$Q = -\tfrac{1}{2} \ln (q^2 + P^2) \tag{6.35b}$$

and

$$F_2(P, q) = q \tan^{-1}(q/P) + P[1 - \frac{1}{2} \ln (q^2 + P^2)]. \tag{6.36}$$

6.5   The transformed Hamiltonian

If the canonical transformation from (q, p) to (Q, P) representation is independent of time and $H(q, p)$ and $K(Q, P)$ are the conservative Hamiltonians of the same flow in each representation, then the numerical value of the Hamiltonian at each point in phase space is independent of the representation, so that

$$K(Q, P) = H(q(Q, P), p(Q, P)). \tag{6.37}$$

This is shown by expressing H as a function of (Q, P) in the following equations:

$$\frac{\partial K}{\partial P} = \dot{Q} = \frac{\partial Q}{\partial q} \dot{q} + \frac{\partial Q}{\partial p} \dot{p}$$

$$= \frac{\partial Q}{\partial q} \frac{\partial H}{\partial p} - \frac{\partial Q}{\partial p} \frac{\partial H}{\partial q}. \tag{6.38}$$

On expressing H in terms of (Q, P) and using the chain rule this becomes

$$\frac{\partial K}{\partial P} = \frac{\partial Q}{\partial q} \left(\frac{\partial H}{\partial Q} \frac{\partial Q}{\partial p} + \frac{\partial H}{\partial P} \frac{\partial P}{\partial p} \right) - \frac{\partial Q}{\partial p} \left(\frac{\partial H}{\partial Q} \frac{\partial Q}{\partial q} + \frac{\partial H}{\partial P} \frac{\partial P}{\partial q} \right)$$

$$= \frac{\partial H}{\partial P} \left(\frac{\partial Q}{\partial q} \frac{\partial P}{\partial p} - \frac{\partial Q}{\partial p} \frac{\partial P}{\partial q} \right), \tag{6.39}$$

and on using the area-preserving property of canonical transformations, equation (6.4a),

$$\frac{\partial K}{\partial P} = \frac{\partial H}{\partial P}. \tag{6.40}$$

Similarly, by considering $\dot{P}$, it may be shown that

$$\frac{\partial K}{\partial Q} = \frac{\partial H}{\partial Q} \tag{6.41}$$

and thus, apart from an irrelevant constant, we have shown that $H = K$.

This derivation, in particular equation (6.38), is valid only for time-independent transformations. For time-dependent transformations the relations between Hamiltonians in different representations is not so simple, as shown in section 6.7.

Example 6.6

·Find the canonical transformation generated by

$$F_1(Q, q) = \lambda q^2 \cot Q, \tag{6.42}$$

λ being a constant. If the Hamiltonian in (q, p) representation is

$$H(q, p) = p^2/2m + \tfrac{1}{2} m\omega^2 q^2, \tag{6.43}$$

find the Hamiltonian in (Q, P) representation. Choose λ to make this Hamiltonian independent of Q and hence find the motion in each representation.

From the generating function

$$p = \frac{\partial F_1}{\partial q} = 2\lambda q \cot Q,$$

$$P = -\frac{\partial F_1}{\partial Q} = \frac{\lambda q^2}{\sin^2 Q} \tag{6.44}$$

or

$$q = \left(\frac{P}{\lambda}\right)^{1/2} \sin Q,$$

$$p = 2(\lambda P)^{1/2} \cos Q. \tag{6.45}$$

Using equation (6.37) we obtain the Hamiltonian in (Q, P) representation:

$$K(Q, P) = P\left(\frac{2\lambda}{m} \cos^2 Q + \frac{m\omega^2}{2\lambda} \sin^2 Q\right). \tag{6.46}$$

This may be made independent of Q by choosing $\lambda = \tfrac{1}{2} m\omega$. Then we have

$$K(Q, P) = \omega P, \tag{6.47}$$

and the equations of motion in (Q, P) representation are:

$$\dot{P} = -\frac{\partial K}{\partial Q} = 0, \quad P = \text{constant};$$

$$\dot{Q} = \frac{\partial K}{\partial P} = \omega = \text{constant}, \quad Q = \omega t + \delta. \tag{6.48}$$

Clearly from equation (6.47) $P = E/\omega$, E being the energy, and in the (q, p) representation the motion is obtained by substituting equations (6.48) into (6.45):

$$q = \frac{1}{\omega} \left(\frac{2E}{m} \right)^{1/2} \sin(\omega t + \delta),$$

(6.49)

$$p = (2mE)^{1/2} \cos(\omega t + \delta).$$

Example 6.7
A canonical transformation

$$Q = f(q, p, \lambda),$$
$$P = g(q, p, \lambda)$$

(6.50)

depends upon a parameter λ. If $F_1(Q, q, \lambda)$ and $F_2(P, q, \lambda)$ are generating functions for this transformation show that $\partial F_1/\partial \lambda = \partial F_2/\partial \lambda$, where q and p are held fixed when differentiating.

We have, from equation (6.26),

$$F_2(P, q, \lambda) = F_1(Q, q, \lambda) + PQ.$$

Differentiating with respect to λ,

$$\frac{\partial F_2}{\partial \lambda} + \frac{\partial F_2}{\partial P} \frac{\partial P}{\partial \lambda} = \frac{\partial F_1}{\partial \lambda} + \frac{\partial F_1}{\partial Q} \frac{\partial Q}{\partial \lambda} + P \frac{\partial Q}{\partial \lambda} + Q \frac{\partial P}{\partial \lambda},$$

(6.51)

but $Q = \partial F_2/\partial P$ and $P = -\partial F_1/\partial Q$, hence

$$\frac{\partial F_2}{\partial \lambda} = \frac{\partial F_1}{\partial \lambda}.$$

(6.52)

This relation is important for the time-dependent theory of section 6.6.

6.6   Time-dependent transformations
The theory of time-dependent transformations, like the theory of time-independent transformations, relies upon the flow being area-preserving.

The transformation

$$Q = f(q, p, t),$$
$$P = g(q, p, t)$$

(6.53)

depending explicitly upon the time, is canonical if, and only if, the Jacobian satisfies

$$\frac{\partial(Q, P)}{\partial(q, p)} = \frac{\partial(f, g)}{\partial(q, p)} = 1,$$

(6.54)

where the time t is held fixed for partial differentiation. The theory of generating functions in section 6.3 may be taken over directly with time introduced as a parameter. Thus the generating functions $F_1(Q, q, t)$ and $F_2(P, q, t)$ generate the transformations

$$p = \frac{\partial F_1}{\partial q}(Q, q, t), \quad P = -\frac{\partial F_1}{\partial Q}(Q, q, t); \tag{6.55}$$

$$p = \frac{\partial F_2}{\partial q}(P, q, t), \quad Q = \frac{\partial F_2}{\partial P}(P, q, t); \tag{6.56}$$

and are related by

$$F_2(P, q, t) = F_1(Q, q, t) + PQ,$$

$$P = -\frac{\partial F_1}{\partial Q}(Q, q, t). \tag{6.57}$$

This may be used to obtain a relation between the time rates of change of the values of F_1 and F_2 at a fixed point in the (q, p) representation. The theory is identical to that of example 6.7. Differentiating equation (6.57) with respect to time and treating q, p and t as the independent variables, as in equation (6.53), gives

$$\frac{\partial F_2}{\partial t} + \frac{\partial F_2}{\partial P}\frac{\partial P}{\partial t} = \frac{\partial F_1}{\partial t} + \frac{\partial F_1}{\partial Q}\frac{\partial Q}{\partial t} + Q\frac{\partial P}{\partial t} + P\frac{\partial Q}{\partial t}. \tag{6.58}$$

Using equations (6.55) and (6.56) we obtain

$$\frac{\partial F_2}{\partial t} = \frac{\partial F_1}{\partial t}. \tag{6.59}$$

Similarly, it can be shown that

$$\frac{\partial F_1}{\partial t} = \frac{\partial F_2}{\partial t} = \frac{\partial F_3}{\partial t} = \frac{\partial F_4}{\partial t}. \tag{6.60}$$

Example 6.8

Show explicitly that $\partial F_1/\partial t = \partial F_2/\partial t$ when $F_1(Q, q, t) = q(Q^2 + t^2)$.

The momentum conjugate to Q is

$$P = -\frac{\partial F_1}{\partial Q} = -2qQ \tag{6.61}$$

and, from equations (6.57) and (6.61),

$$F_2(P, q, t) = q(Q^2 + t^2) + PQ,$$

$$Q = -P/2q. \tag{6.62}$$

Hence

$$F_2(P, q, t) = q\,t^2 - P^2/4q, \tag{6.63}$$

and consequently

$$\frac{\partial F_2}{\partial t} = 2qt = \frac{\partial F_1}{\partial t} . \tag{6.64}$$

6.7 Hamiltonians under time-dependent transformations

For time-independent transformations the Hamiltonians are simply related by equation (6.37): the value of the Hamiltonian is independent of the representation. This is not true for time-dependent transformations. For example, a transformation from an inertial to an accelerating reference frame introduces forces additional to those present in the inertial frame and these forces necessitate an addition to the Hamiltonian.

The Hamiltonian in the new representation is obtained by direct differentiation along the flow, with q, p and t as independent variables:

$$\dot{Q} = \frac{\partial Q}{\partial q} \dot{q} + \frac{\partial Q}{\partial p} \dot{p} + \frac{\partial Q}{\partial t}. \tag{6.65}$$

The first two terms on the right-hand side of this equation are those obtained in the time-independent theory, equation (6.38), so, on repeating the analysis of section 6.5, we obtain

$$\dot{Q} = \frac{\partial H}{\partial P} + \frac{\partial Q}{\partial t}, \tag{6.66}$$

where the last term is new. Since $Q = \partial F_2/\partial P$, we have

$$\frac{\partial Q}{\partial t} = \frac{\partial}{\partial t} \left(\frac{\partial F_2}{\partial P} \right) = \frac{\partial}{\partial P} \left(\frac{\partial F_2}{\partial t} \right) . \tag{6.67}$$

Thus, equation (6.66) may be written

$$\dot{Q} = \frac{\partial}{\partial P} \left(H + \frac{\partial F_2}{\partial t} \right), \tag{6.68}$$

and, similarly

$$\dot{P} = - \frac{\partial}{\partial Q} \left(H + \frac{\partial F_1}{\partial t} \right). \tag{6.69}$$

Using equation (6.60), we can replace F_1 by F_2 in this equation. Thus the Hamiltonian in the (Q, P) representation is

$$K(Q, P, t) = H(q, p, t) + \frac{\partial F_2}{\partial t} (P, q, t),$$

where q and p are expressed in terms of Q and P. Because of the relation (6.60), any of the four generating functions may be used in forming the new Hamiltonian, and we have

$$K(Q, P, t) = H(q, p, t) + \frac{\partial F_j}{\partial t} \quad (j = 1, 2, 3, 4). \tag{6.70}$$

Example 6.9

The transformation from a fixed to a moving reference frame is

$$Q = q - D(t), \tag{6.71}$$

$D(t)$ being the distance between the origins at time t. Find the full transformation, the Hamiltonian in the (Q, P) representation corresponding to $p^2/2m + V(q)$ in the (q, p) representation, and the equations of motion.

A generating function for the transformation (6.71) is

$$F_2(P, q, t) = P[q - D(t)] \tag{6.72}$$

and so

$$p = \frac{\partial F_2}{\partial q} = P.$$

The transformed Hamiltonian is then

$$K(Q, P, t) = P^2/2m + V(Q + D) - P\dot{D} \tag{6.73}$$

and the equations of motion are

$$\dot{Q} = \frac{\partial K}{\partial P} = P/m - \dot{D} \tag{6.74}$$

$$\dot{P} = -\frac{\partial K}{\partial Q} = -\frac{\partial V}{\partial Q}(Q + D)$$

or

$$m\ddot{Q} = -\frac{\partial V}{\partial Q}(Q + D) - m\ddot{D}. \tag{6.75}$$

The $m\ddot{D}$ term is the force due to the acceleration of the reference frame. For uniform motion this is zero.

6.8 Group property and infinitesimal canonical transformation

The inverse of a canonical transformation and the transformation produced by two successive canonical transformations must be area-preserving and hence must be canonical transformations. Canonical transformations are associative and the identity transformation is canonical. Consequently the set of all canonical transformations forms a group.

Of particular importance are families of canonical transformations close to the identity. The generating function may be written in the form

$$F_2(P, q) = qP + \epsilon W(q, P, \epsilon), \tag{6.76}$$

with small ϵ; this gives the identity at $\epsilon = 0$. The transformation is

$$p = P + \epsilon \, \frac{\partial W}{\partial q} \, (q, P, \epsilon),$$

$$Q = q + \epsilon \, \frac{\partial W}{\partial P} \, (q, P, \epsilon),$$

(6.77)

which, to first order in ϵ, may be written in the explicit form:

$$Q = q + \epsilon \, \frac{\partial G}{\partial p} \, (q, p),$$

$$P = p - \epsilon \, \frac{\partial G}{\partial q} \, (q, p),$$

(6.78)

where $G(q, p) = \underset{\epsilon \to 0}{\mathrm{Lim}} \, W(q, p, \epsilon)$ is called the *generator* of the *infinitesimal canonical transformation* (6.78). Note that there is a distinction between the generator, G, and the generating function F_2.

We may regard (Q, P) as functions of ϵ; then, on rearranging equation (6.78) and taking the limit as $\epsilon \to 0$, we obtain the differential equations:

$$\frac{\partial Q}{\partial \epsilon} = \frac{\partial G}{\partial p} \, (q, p) = \frac{\partial G}{\partial P} \, (Q, P);$$

$$\frac{\partial P}{\partial \epsilon} = - \, \frac{\partial G}{\partial q} \, (q, p) = - \, \frac{\partial G}{\partial Q} \, (Q, P);$$

since $(Q, P) \to (q, p)$ as $\epsilon \to 0$. These are of Hamiltonian form. In particular, choosing G to be the Hamiltonian, $G = H(q, p)$, we see that the flow taking a phase point $(q(t), p(t))$ to a nearby point $(q(t + \delta t), p(t + \delta t))$ may be regarded as an infinitesimal canonical transformation whose generator is the Hamiltonian.

Example 6.10
The generator $G(q, p) = q$ gives

$$Q = q,$$
$$P = p - \epsilon,$$

which is an infinitesimal translation along the momentum axis.

Exercises for chapter 6

(1) Find the canonical transformation produced by the following generating functions:

(a) $F_1(Q, q) = q\, e^Q$;

(b) $F_2(P, q) = P \tan q$.

(2) Determine which of the following functions generate canonical transformations, and find the transformation for those that do:

(a) $F_1(Q, q) = \frac{1}{2} q^2 \tan Q$;

(b) $F_1(Q, q) = q^2 + Q^4$.

(3) If k is a constant and t the time, determine which of the following are canonical transformations:

(a) $Q = \frac{1}{2} q^2$, $P = p/q$;

(b) $Q = \tan q$, $P = (p - k) \cos^2 q$;

(c) $Q = \sin q$, $P = (p - k)/\cos q$;

(d) $Q = q^{\frac{1}{2}} e^t \cos p$, $P = q^{\frac{1}{2}} e^{-t} \sin p$.

(4) The new coordinate Q is given explicitly in terms of the old coordinate q by

$$Q = f(q, t).$$

Write down the most general $F_2(P, q, t)$ generating function and obtain an expression for the momentum conjugate to Q.

(5) The old coordinate q is given explicitly in terms of the new coordinate Q by

$$q = f(Q, t).$$

Write down the most general $F_3(Q, p, t)$ generating function and obtain an expression for the momentum conjugate to Q.

(6) Show that the transformation

$$Q = q \cos \alpha - p \sin \alpha,$$

$$P = q \sin \alpha + p \cos \alpha,$$

α being a constant, is canonical. Show how the unit square in the (q, p) representation is transformed if the transformation is interpreted as moving the phase point, as in figure 6.1a. Also show the (Q, P) co-ordinate axes in the (q, p) phase diagram if the transformation is interpreted as changing the coordinate system.

(7) For what values of α and β is the following a canonical transformation:

$$q = P^\alpha \cos \beta Q, \quad p = P^\alpha \sin \beta Q \quad (P \geqslant 0)\,?$$

In this case sketch the (Q, P) coordinate system in the (q, p) phase diagram.

(8) Show that the generating functions F_1, F_2, F_3 and F_4 are related by

$$F_1(Q, q) = F_2(P, q) - PQ, \quad Q = \partial F_2/\partial P,$$
$$F_3(Q, p) = F_1(Q, q) - pq, \quad p = \partial F_1/\partial q,$$
$$F_4(P, p) = F_2(P, q) - pq, \quad p = \partial F_2/\partial q.$$

For the F_3 and F_4 generating functions determine the equations giving the transformation.

(9) Find F_1, F_3 and F_4 if

$$F_2(P, q) = q^2 e^P.$$

(10) The coordinates of a canonical transformation $(q, p) \rightarrow (Q, P)$ are related by $Q = q^2$. Find a conjugate momentum P, and the Hamiltonian in the (Q, P) representation corresponding to

$$H(q, p) = p^2 + q^2.$$

Find the time dependence of (Q, P) and, hence, (q, p).

(11) A new coordinate Q is given by

$$Q = \tan^{-1}(\lambda q/p),$$

where λ is a constant. Show that the momentum conjugate to Q is

$$P = (p^2 + \lambda^2 q^2)/2\lambda + f(\lambda q/p, t),$$

f being an arbitrary function. Apply this transformation to the linear oscillator with Hamiltonian

$$H = p^2/2m + \tfrac{1}{2} m\omega^2 q^2,$$

choosing λ and f to simplify the problem. Solve for (Q, P), and hence (q, p), in terms of the time.

(12) Find an $F_2(P, q)$ generating function giving

$$q = Q + \epsilon QP$$

and the identity transformation at $\epsilon = 0$. Find the equivalent $F_1(Q, q)$ generating function and examine its limit as $\epsilon \rightarrow 0$.

(13) The Hamiltonian of a freely falling body is

$$H(z, p) = p^2/2m + mgz,$$

z being the height of the particle above the ground and m its mass. By expressing z as a function of (p, P) and using an F_4 generating function find a canonical transformation so that the new Hamiltonian is

$$K(Q, P) = P.$$

Solve for $(Q(t), P(t))$ and hence find $(q(t), p(t))$.

(14) Show that if a time-independent transformation $(q, p) \rightarrow (Q, P)$ has the Jacobian

$$\frac{\partial(Q,P)}{\partial(q,p)} = \text{constant} = \lambda \neq 0,$$

then the flow in the (Q, P) representation is area-preserving and has Hamiltonian

$$K(Q,P) = \lambda H(q(Q,P), p(Q,P)),$$

where $H(q,p)$ is the time-independent Hamiltonian of the flow in the (q,p) representation. If $\bar{Q} = Q/\lambda$, $\bar{P} = P$ show that the transformation $(q,p) \rightarrow (\bar{Q}, \bar{P})$ is canonical.

(15) Find the canonical transformation generated by

$$F_1(Q, q, t) = \tfrac{1}{2} m \, \omega(t) q^2 \cot Q,$$

where $\omega(t)$ is a given function of time. Determine the equations of motion in the (Q, P) representation for the linear oscillator with Hamiltonian

$$H(q, p, t) = p^2/2m + \tfrac{1}{2} m\omega(t)^2 q^2.$$

(16) If

$$F_1(Q, q) = \lambda q^2 \cot Q,$$

find $F_2(P, q)$ and show explicitly that

$$\partial F_1/\partial \lambda = \partial F_2/\partial \lambda.$$

(17) Using the generating function

$$F_2(P, q) = Pq + \epsilon a q^3 P + \epsilon b q P^3,$$

where ϵ is small and a and b are constants, show that

$$q = Q - 3\epsilon bQP^2 - \epsilon aQ^3 + O(\epsilon^2),$$
$$p = P + \epsilon bP^3 + 3\epsilon aQ^2P + O(\epsilon^2).$$

Find the values of a and b which reduce the anharmonic oscillator

$$H(q,p) = \tfrac{1}{2} (p^2 + \omega^2 q^2) + \epsilon \beta q^4$$

to the form

$$K(Q,P) = \tfrac{1}{2}(P^2 + \omega^2 Q^2) + \epsilon c(P^2 + \omega^2 Q^2)^2 + O(\epsilon^2),$$

c being a constant to be found. Hence show that, to this order in ϵ,

$$Q = A\cos(\Omega t + \delta), P = \omega A \sin(\Omega t + \delta), \Omega = \omega + 3\epsilon\beta A^2/2\omega,$$

where A and δ are constants. Hence find $q(t)$ to this order in ϵ.

(18) For a conservative system with Hamiltonian $H(q, p)$, show that the special generating function for the transformation $(q_1, p_1) \rightarrow (q_2, p_2)$, where $q_i = q(t_i)$, $p_i = p(t_i)$ are points on a phase curve of energy E, $H(q, p) = E$, is given by

$$W(q_2, q_1; E) = \int_{q_1}^{q_2} p(q, E) \, dq.$$

Further, show that

$$\frac{\partial W}{\partial E} = t_2 - t_1,$$

Find W explicitly in the following cases:

(a) free motion;
(b) linear potential;
(c) linear force.

(19) Show that the generator $G(q, p) = -ap$, where a is a positive constant, generates a linear transformation in the direction of the generalized coordinate.

(20) Determine the generator which produces infinitesimal rotation in the phase plane.

(21) Determine a generating function for the time-dependent transformation

$$q = Q + Pt/m, \quad p = P.$$

Show that the Hamiltonian

$$H(q, p) = p^2/2m + V(q)$$

transforms to

$$K(Q, P, t) = V(Q + \frac{Pt}{m}).$$

Consider the particular case $V(q) = -q$; write down and solve Hamilton's equations in the (Q, P) representation and from this find the motion in the (q, p) representation.

7 ANGLE-ACTION VARIABLES

7.1 The simplest variables

The variables used to formulate a problem are not usually the best for solving it. Further, we often want to use the solution of a relatively simple system as a starting point for the solution of a more complex system as, for example, in chapters 8 and 9. Then it is best to choose variables so that the solution of the simpler problem is expressed in the simplest possible way: for the bounded motion of conservative Hamiltonian systems these variables are called *angle–action variables*.

For systems of one degree of freedom such motion is usually periodic in time and, as we have seen in chapter 4, may be either a rotation or a libration. The theory of this chapter is valid in both cases but, since there are slight differences, we first consider libration. The theory requires that a phase curve should be closed, but not a separatrix. To be specific, we choose a Hamiltonian of the form

$$H(q, p) = p^2/2m + V(q) \tag{7.1}$$

with typical potential and phase curves as shown in figure 7.1.

In the (q, p) representation the phase curve of energy E is represented by the two-valued function

$$p(q, E) = \pm [2m (E - V(q))]^{\frac{1}{2}}. \tag{7.2}$$

The multivaluedness of this is unsatisfactory. Therefore we seek a new pair of conjugate variables (θ, I) with the properties:

(P1) Each phase curve is labelled uniquely by I, which is a constant along every phase curve.

(P2) Each point on a phase curve is labelled by a single-valued function of θ.

In the (θ, I) representation the contours are lines of constant I, so that the Hamiltonian is independent of θ. This is expressed in Hamilton's equation of motion

$$\dot{I} = 0 = -\frac{\partial H}{\partial \theta}. \tag{7.3}$$

Fig. 7.1 Simple potential and phase curves

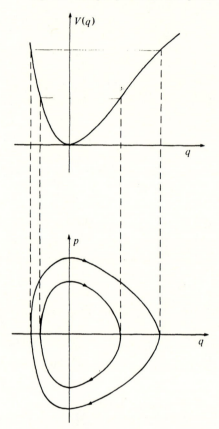

Since I is constant, $\partial H / \partial I$, which is a function of I only, is also constant. Then the other equation of motion,

$$\dot{\theta} = \frac{\partial H}{\partial I} = \text{constant},\qquad (7.4)$$

shows that θ increases linearly with time.

The angle-action variables are obtained by making θ increase by 2π in each period. Then θ is named the *angle variable*, and I the *action variable*. The phase curves in the angle-action representation are thus straight lines parallel to the θ-axis as shown below

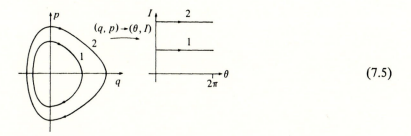

(7.5)

The time dependence of θ is given directly by integrating equation (7.4)

$$\theta = \omega(I)t + \delta, \quad \omega(I) = \partial H/\partial I, \tag{7.6}$$

where δ is an arbitrary constant. Since θ increases by 2π during one period T, ω is the angular frequency of the motion

$$\omega(I) = 2\pi/T = \partial H/\partial I. \tag{7.7}$$

Further, since the points (θ, I) and $(\theta + 2\pi, I)$ label the same phase point, the (θ, I) coordinates can be represented upon a semi-infinite cylinder where θ is the angle around the cylinder and I the coordinate along its axis,

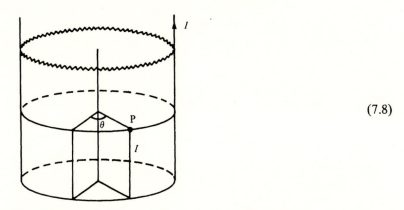

(7.8)

On this cylinder both q and p are periodic functions of θ with period 2π:

$$q(\theta + 2\pi, I) = q(\theta, I)$$
$$p(\theta + 2\pi, I) = p(\theta, I). \tag{7.9}$$

A separatrix divides phase space into invariant regions containing phase ·curves with different properties. Within each such invariant region either all motions are periodic or none are. When the motion is periodic, angle–action variables may be defined, but they are not defined on a separatrix, nor are angle–action variables in different regions simply related.

For instance, the example discussed in section 4.5 has four separatrixes, and four invariant regions of which three contain periodic motion with different angle-action variables.

Example 7.1

The transformation of example 6.6,

$$q = \left(\frac{2I}{m\omega} \right)^{\frac{1}{2}} \sin \theta, \quad p = (2Im\omega)^{\frac{1}{2}} \cos \theta$$

transforms the Hamiltonian

$$H(q,p) = p^2/2m + \tfrac{1}{2} m\omega^2 q^2$$

into

$$K(I) = \omega I.$$

7.2 The Hamiltonian in angle-action representation

We confine our attention to Hamiltonians of the form (7.1). The Hamiltonian in angle-action representation is obtained by comparing the area enclosed by a phase curve during one whole period in one of the two representations with the corresponding area in the other. Since the transformation $(q, p) \rightarrow (\theta, I)$ is canonical, the theory of chapter 6 shows that these areas are the same. In the (q, p) representation the area enclosed by the phase curve of energy E is

$$A(E) = \oint dq \, p(q, E) \tag{7.10}$$

$$= 2 \int_{q_1}^{q_2} dq \, [2m(E - V(q))]^{\frac{1}{2}}, \tag{7.11}$$

where $V(q_i) = E$, $i = 1, 2$. In the (θ, I) representation the area is

$$A(E) = \int_0^{2\pi} d\theta \, I = 2\pi I. \tag{7.12}$$

Thus for libration the action variable is proportional to the area; this important property is sometimes used to *define* the action variable.

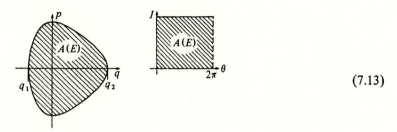

$$(7.13)$$

By comparing these two expressions for the area we obtain the action variable as a function of energy:

$$I(E) = \frac{1}{\pi} \int_{q_1}^{q_2} \mathrm{d}q \ [2m \ (E - V(q))]^{\frac{1}{2}}, \qquad (7.14)$$

which is a monotonic function of E within each invariant region.

By inverting (7.14) we obtain the energy $E(I)$ as a function of the action, as shown in example 7.2 below. We usually denote this function by $H(I)$ or $K(I)$, reserving the symbol E for a particular value of the energy.

The action variable has the same dimensions as angular momentum or as energy $\times$ time. The angle variable is dimensionless and it is sometimes, though not always, an angle in configuration space.

For potentials of the type shown in figure 7.1 there is a minimum energy, E_0 say, below which motion is impossible; for the potential of figure 7.1, $E_0 = 0$. At this energy the phase curves in the (q, p) representation reduce to a point and $I(E_0) = 0$; that is, the action has a natural boundary at $I = 0$.

Example 7.2

Find the action and Hamiltonian for a particle moving in the potential

$$V(q) = U \tan^2 \alpha q, \qquad (7.15)$$

where U and α are positive constants.

Equation (7.14) gives the action as

$$I = \frac{1}{\pi} \int_{q_1}^{q_2} \mathrm{d}q \ [2m \ (E - U \tan^2 \alpha q)]^{\frac{1}{2}}, \qquad (7.16)$$

where $\tan^2 \alpha q_2 = E/U$, $q_1 = -q_2$. Evaluation of this integral gives

$$\alpha I = [2m \ (E + U)]^{\frac{1}{2}} - [2m \ U]^{\frac{1}{2}} \qquad (7.17)$$

The potential and phase curves in (q, p) and (θ, I) representation are shown in figure 7.2.

Fig. 7.2 The $\tan^2 \alpha q$ potential and phase curves in the (q, p) and (θ, I) representations.

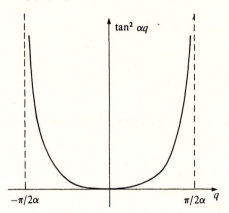

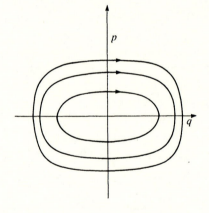
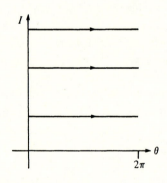

Note that $I \geqslant 0$ with equality only at $E = 0$. Inverting this relation gives the Hamiltonian,

$$H(\theta, I) = \alpha I \left[\alpha I + 2\,(2mU)^{\frac{1}{2}}\right]/2m. \tag{7.18}$$

The angular frequency is then given by

$$\omega = \partial H/\partial I = \alpha[\alpha I + (2mU)^{\frac{1}{2}}]/m$$

$$= \alpha[2\,(E + U)/m]^{\frac{1}{2}}. \tag{7.19}$$

Example 7.3

Find the action for a ball of mass m bouncing elastically between two fixed walls at a distance d apart, with a speed v.

This example cannot be described by a continuous Hamiltonian since at each collision the momentum changes discontinuously; however the motion is close to that of the continuous Hamiltonians of example 7.2 and exercise 7.1. Further, because the motion is particularly simple, it may be used to help understand more complex systems, as in section 9.2.

In between collisions with the walls, the momentum is constant. At each collision the momentum changes sign instantaneously. The phase curve for the ball of energy E is thus the rectangle shown below.

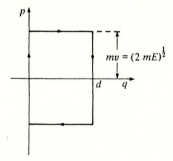

The action is thus

$$I = \frac{1}{2\pi}\ (\text{area of rectangle}) = \frac{1}{2\pi}\ \times 2(2mE)^{\frac{1}{2}} \times d$$

$$= d(2mE)^{\frac{1}{2}}/\pi = mvd/\pi, \tag{7.20}$$

v being the speed of the ball. The energy, as a function of the action, is

$$E = (\pi I/d)^2/2m. \tag{7.21}$$

Note that, for large actions, this approaches the Hamiltonian (7.18), if we put $d = \pi/\alpha$.

7.3   The dependence of the angle variable upon *q*

For fixed action, I, the relation between q and θ is obtained by considering the area between segments of two adjacent phase curves of action $I + \delta I$ and I, as illustrated in figure 7.3 below.

Fig. 7.3 Segments of two phase curves of action I and $I + \delta I$ in the (q, p) and (θ, I) representations. The shaded area S in the (q, p) representation is transformed into the shaded area in the (θ, I) representation. This is the region used to determine $\theta(q)$.

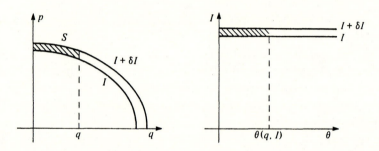

Let δA be the area of the shaded segment S in the (q, p) plane; by definition this equals the area of the shaded segment in the (θ, I) plane. In the (q, p) plane we have

$$\delta A = \iint_S dq\, dp$$

$$= \int_0^q dq\, [p(q, I + \delta I) - p(q, I)]$$

$$= \delta I \int_0^q dq\, \frac{\partial p}{\partial I}(q, I) + O(\delta I)^2, \qquad (7.22)$$

where $p(q, I)$ is obtained from $p(q, E)$, equation (7.2), by substituting $E(I)$ for E. In the (θ, I) plane, the area is

$$\delta A = \delta I\, \theta(q, I) + O(\delta I)^2 . \qquad (7.23)$$

Equating these two expressions for δA and taking the limit as $\delta I \to 0$,

$$\theta(q) = \int_0^q dq\, \frac{\partial}{\partial I}\, p(q, I)$$

$$= \frac{\partial}{\partial I} \int_0^q dq\, p(q, I). \qquad (7.24)$$

In deriving this relation we arbitrarily choose one boundary of the shaded region to be the p and I axes: other choices give angle variables differing from that of equation (7.24) by unimportant additive constants.

In evaluating $\theta(q)$ from equation (7.24) care must be taken because $p(q, I)$ is a multivalued function of q. For the phase curves illustrated in figures 7.1 and 7.3, p is positive when q is increasing, otherwise it is negative. Thus $\theta(q)$ increases monotonically and continuously as the phase point moves clockwise around the phase curve, and in one revolution θ increases by 2π. Typical graphs of $\theta(q)$ and its inverse $q(\theta)$ are shown in figure 7.4.

Fig. 7.4 Figure showing typical behaviour of $\theta(q)$ and $q(\theta)$.

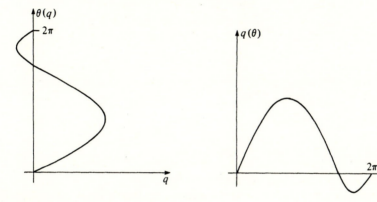

Example 7.4

Find the angle variable for the potential

$$V(q) = U \tan^2 \alpha q. \tag{7.25}$$

According to equation (7.24) the angle variable is given by

$$\theta(q) = \frac{\partial}{\partial I} \int_0^q dq \, [2m \, (E(I) - U \tan^2 \alpha q)]^{\frac{1}{2}}, \tag{7.26}$$

where $E(I)$ is given in example 7.2, equation (7.18). On taking the derivative inside the integral (usually the easiest way to do these integrals) and using equation (7.19) we obtain, using appendix 2 (equation A2.9),

$$\theta(q) = m \, \frac{dE}{dI} \int_0^q dq \, [2m \, (E - U \tan^2 \alpha q)]^{-\frac{1}{2}}$$

$$= \sin^{-1} \cdot \left[\left(\frac{E+U}{E} \right)^{\frac{1}{2}} \sin \alpha q \right], \tag{7.27}$$

where $\omega = \mathrm{d}E/\mathrm{d}I$ is the angular frequency. The inverse function $q(\theta)$ is thus

$$q(\theta) = \frac{1}{\alpha} \sin^{-1} \left[\left(\frac{E}{E+U} \right)^{\frac{1}{2}} \sin \theta \right]. \tag{7.28}$$

Putting $\theta = \omega t + \delta$, equation (7.6), gives q as a function of time.

7.4   Generating functions

Angle-action variables are special, so we use a special symbol S instead of F to denote a generating function for the canonical transformation $(q, p) \to (\theta, I)$. The generating function $S_2(I, q)$ is simply

$$S_2(I, q), = \int_0^q \mathrm{d}q \, p(q, I), \tag{7.29}$$

which is a multivalued function of q. To show that this is the generating function, differentiate with respect to q giving

$$p = \frac{\partial S_2}{\partial q} (I, q). \tag{7.30}$$

By using (7.24) we find that

$$\theta = \frac{\partial S_2}{\partial I} (I, q), \tag{7.31}$$

so that $S_2(I, q)$ is the appropriate generating function.

An important result, of use later, concerns the change in the value of S_2 and the associated generating function $S_1(\theta, q)$ during one period of the motion. The change in S_2 is

$$\Delta S_2(I) = \oint \mathrm{d}q \, \frac{\partial S_2}{\partial q}$$

$$= \oint \mathrm{d}q \, p = 2\pi I. \tag{7.32}$$

The generating function $S_1(\theta, q)$ is given by

$$S_1(\theta, q) = S_2(I, q) - \theta I,$$

$$\theta = \partial S_2 / \partial I \tag{7.33}$$

and its change in one period is

$$\Delta S_1 = \Delta S_2 - \Delta(I\theta)$$

$$= \Delta S_2 - I\Delta\theta = 0. \tag{7.34}$$

Thus $S_1(\theta, q)$ returns to its initial value after one period whilst $S_2(I, q)$ increases by $2\pi I$, so that $S_1(\theta, q)$ is periodic in θ whilst $S_2(I, q)$ is not.

Example 7.5
Find the angle-action variables and the generating functions $S_1(\theta, q)$ and $S_2(I, q)$ for the linear oscillator with the Hamiltonian

$$H(q, p) = p^2/2m + \tfrac{1}{2} m\omega^2 q^2. \tag{7.35}$$

Using equation (7.14), the action is

$$I = \frac{1}{\pi} \int_{-q_1}^{q_1} dq \, [2m(E - \tfrac{1}{2} m\omega^2 q^2)]^{\frac{1}{2}} \qquad (m\omega^2 q_1^2 = 2E)$$

$$= E/\omega, \tag{7.36}$$

so that the Hamiltonian in angle-action representation is

$$H(I) = \omega I, \tag{7.37}$$

and the frequency of the motion is $\partial H/\partial I = \omega$.

By equating the two Hamiltonians (7.37) and (7.35) we obtain p in terms of q and I,

$$p = [2m\omega I - (m\omega q)^2]^{\frac{1}{2}}. \tag{7.38}$$

Then on using equation (7.24) we can find the angle variable,

$$\theta = \int_0^q dq \left[\frac{m\omega}{2I - m\omega q^2} \right]^{\frac{1}{2}} = \sin^{-1} \left[q \left(\frac{m\omega}{2I} \right)^{\frac{1}{2}} \right], \tag{7.39}$$

or

$$q = \left(\frac{2I}{m\omega} \right)^{\frac{1}{2}} \sin\theta. \tag{7.40}$$

The $S_2(I, q)$ generating function is obtained from (7.29):

$$S_2(I, q) = \int_0^q dq \, (2m\omega I - m^2\omega^2 q^2)^{\frac{1}{2}}$$

$$= I \sin^{-1} \left[q \left(\frac{m\omega}{2I} \right)^{\frac{1}{2}} \right] + \frac{1}{2} q \, (2Im\omega - m^2\omega^2 q^2)^{\frac{1}{2}} \tag{7.41}$$

and it may be verified that (7.39) is obtained from this by differentiating with respect to I. Using (7.33) we find

$$S_1(\theta, q) = \tfrac{1}{2} m\omega q^2 \cot\theta, \tag{7.42}$$

from which we find p in terms of the angle–action variables

$$p = \partial S_1 / \partial q = m\omega q \cot \theta$$

$$= (2Im\omega)^{\frac{1}{2}} \cos \theta. \tag{7.43}$$

It will be noticed that the derivation (7.39) is similar to the method used to obtain the implicit relation $q(t)$ in equation (4.51). This is a consequence of the linear relation between θ and t.

7.5 Rotations

For rotations the phase curves on the plane are not closed and the co-ordinate $\psi(t)$ is either a continuously increasing, or a continuously decreasing, function of time. The Hamiltonian is periodic in ψ, and we suppose this period to be 2π:

$$H(\psi + 2\pi, p) = H(\psi, p). \tag{7.44}$$

As before, we seek a pair of angle–action variables (θ, I) such that each phase curve is labelled by an action I and each point on a phase curve is labelled by a variable θ that is linear in time, as in equation (7.6). Then the phase curves in each representation look like those depicted in (7.45):

$$\tag{7.45}$$

The action is given as a function of energy by equating the areas between the phase curves and horizontal axes in the (ψ, p) and (θ, I) representations:

$$I(E) = \frac{1}{2\pi} \int_0^{2\pi} d\psi \, p(\psi, E), \tag{7.46}$$

where $p(\psi, E)$ is a function of ψ given by the solution to the energy equation

$$H(\psi, p) = E. \tag{7.47}$$

Usually this has more than one solution. When the Hamiltonian has the form in equation (7.1) there are two solutions, corresponding to the different signs of the square root and representing rotations in opposite directions. Thus for each energy there are, in this case, two actions I defined by equation (7.46), corresponding to each direction of motion.

Different choices of horizontal axis give actions differing from equation (7.46) by additive constants. Thus, unlike librations, for rotations there is no natural boundary for the action, and an arbitrary constant may be added to it.

The form of the equations for $\theta(\psi)$ and the generating functions are identical to those found above, with ψ substituted for q, but, since $p(\psi, I)$ is single-valued for rotations, $\theta(\psi)$ is a continuously increasing or decreasing function of ψ.

Example 7.6
Find the action variable for a body freely rotating about an axis.
From section 4.6, equation (4.36), the Hamiltonian is

$$H(\psi, p) = p^2/2G, \tag{7.48}$$

where G is the moment of inertia. By convention I is chosen to have the same sign as p, the angular momentum,

$$I = \frac{\pm 1}{2\pi} \int_0^{2\pi} d\psi \, (2GE)^{\frac{1}{2}} \; = \; \pm (2GE)^{\frac{1}{2}}. \tag{7.49}$$

In this special case, where the potential is equal to zero, the action variable is equal to the angular momentum of the body rotating with angular frequency $\omega = \partial E/\partial I = I/G$.

Example 7.7
Find the angle–action variables of the rotations of a body moving in the periodic potential

$$V(\psi) = \begin{cases} -U\psi/\pi & (-\pi \leqslant \psi \leqslant 0) \\ U\psi/\pi & (0 \leqslant \psi \leqslant \pi) \end{cases} \tag{7.50a}$$

$$V(\psi) = V(\psi + 2\pi).$$

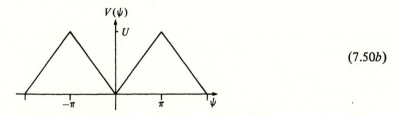

$$\tag{7.50b}$$

For rotations the energy, E, must be greater than U, otherwise the system librates. Then the action variable is obtained from equation (7.46)

$$I = \frac{(2G)^{\frac{1}{2}}}{2\pi} \int_{-\pi}^{0} d\psi \, \left(E + \frac{U\psi}{\pi} \right)^{\frac{1}{2}} + \frac{(2G)^{\frac{1}{2}}}{2\pi} \int_{0}^{\pi} d\psi \, \left(E - \frac{U\psi}{\pi} \right)^{\frac{1}{2}}$$

$$= \frac{(2G)^{\frac{1}{2}}}{\pi} \int_{0}^{\pi} d\psi \, \left(E - \frac{U\psi}{\pi} \right)^{\frac{1}{2}}$$

$$= \frac{2(2G)^{\frac{1}{2}}}{3U} \; [E^{\frac{3}{2}} - (E-U)^{\frac{3}{2}}]. \tag{7.51}$$

By differentiating with respect to E we obtain the frequency

$$\frac{1}{\omega} = \frac{\mathrm{d}I}{\mathrm{d}E} = \frac{(2G)^{\frac{1}{2}}}{U} [E^{\frac{1}{2}} - (E-U)^{\frac{1}{2}}]$$

or

$$\omega = \frac{E^{\frac{1}{2}} + (E-U)^{\frac{1}{2}}}{(2G)^{\frac{1}{2}}}. \tag{7.52}$$

The angle variable is given by (7.24):

$$\theta(\psi) = \int_0^\psi \mathrm{d}\psi \, \frac{\partial p}{\partial I} = \omega \int_0^\psi \mathrm{d}\psi \, \frac{\partial p}{\partial E}$$

$$= \omega \left(\frac{G}{2}\right)^{\frac{1}{2}} \int_0^\psi \frac{\mathrm{d}\psi}{[E - V(\psi)]^{\frac{1}{2}}}, \tag{7.53}$$

which gives

$$\theta(\psi) = \begin{cases} \dfrac{\pi \{[1 + (\bar{U}\psi/\pi)]^{\frac{1}{2}} - 1\}}{1 - (1 - \bar{U})^{\frac{1}{2}}} & (-\pi \leqslant \psi \leqslant 0) \\[4mm] \dfrac{\pi \{1 - [1 - (\bar{U}\psi/\pi)]^{\frac{1}{2}}\}}{1 - (1 - \bar{U})^{\frac{1}{2}}} & (0 \leqslant \psi \leqslant \pi), \end{cases} \tag{7.54}$$

where $\bar{U} = U/E$. A sketch of the graph of $\theta(\psi)$ for various $\bar{U}$ is shown below:

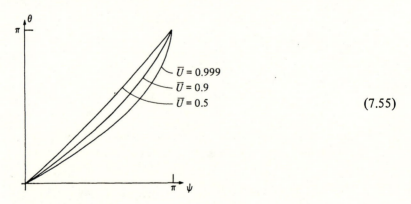

$$\tag{7.55}$$

Exercises for chapter 7
Appendix 2 contains some helpful integrals.
(1) Sketch the phase curves for the motion of a particle of mass m moving in the potential

$$V(q) = A \left(\frac{q}{d} \right)^{2n},$$

A and d being positive constants and n a positive integer. Show that the action I and energy E are related by

$$E = \left(\frac{n\pi I}{dJ_n} \right)^{\frac{2n}{1+n}} \left(\frac{1}{2m} \right)^{\frac{n}{1+n}} A^{\frac{1}{1+n}},$$

where

$$J_n = \int_0^1 dx \, (1-x)^{\frac{1}{2}} x^{-1+1/2n}.$$

For large n, $J_n = 2n + O(n^{-1})$. Show that, in this limit, equation (7.21) is regained.

(2) Show that the area enclosed by the separatrix of the vertical pendulum with Hamiltonian

$$H = \tfrac{1}{2} p^2 - \alpha^2 \cos \psi$$

is 16α. Deduce that the maximum value of the action for librating motion is $8\alpha/\pi$.

(3) Sometimes the action variable is *defined* to be equal to the area under the phase curve. Show that, with this definition, the conjugate variable increases by unity during each period.

(4) Show that the potential

$$V(r) = \epsilon \left(1 - \frac{d}{r} \right)^2 \qquad (\epsilon > 0, \; d > 0)$$

supports vibrations for energies satisfying $0 < E < \epsilon$, and that there is a separatrix of energy ϵ. Also show that, where they exist, the energy and action are related by

$$E = \epsilon \left[1 - \left(\frac{A}{I+A} \right)^2 \right] \qquad A = d\sqrt{2m\epsilon}.$$

Over what values does the action vary?

Show that the period of the librating motion for energies close to the separatrix energy is large, and tends to infinity as E approaches ϵ.

(5) Consider the motion of a particle of mass m in the potential

$$V(r) = \tfrac{1}{2} (kr^2 + \alpha/r^2) \qquad (k > 0, \; \alpha > 0, \; r > 0).$$

Show that the energy and action are related by

$$E = E_0 + 2I \sqrt{k/m} \qquad (E_0 = \sqrt{k\alpha})$$

and deduce that the frequency of the motion is independent of the amplitude and is

$\omega = 2\sqrt{k/m}.$

Show further that r is given in terms of the angle variable, θ, by

$r^2 k = E + (E^2 - k\alpha)^{\frac{1}{2}} \sin\theta.$

You may assume that

$$\int_{x_1}^{x_2} dx (-Ax^2 + 2Bx - C)^{\frac{1}{2}} x^{-1} = \pi(BA^{-\frac{1}{2}} - C^{\frac{1}{2}})$$

$$(A > 0, B > 0, C > 0),$$

x_i being the zeros of the integrand.

(6) Find the angle–action variables for a particle of mass m rotating in the periodic potential

$V(\psi) = A\psi \quad (0 \leqslant \psi \leqslant \alpha < \pi)$

$\quad\quad\quad = A\alpha \quad (\alpha \leqslant \psi \leqslant \pi),$

$V(-\psi) = V(\psi),$

where A and α are positive constants. Determine the frequency of the motion of energy E.

(7) A particle of mass m is attached to the origin by a flexible light spring of length d so that it moves freely if its distance from the origin is less than d, but otherwise moves under the action of the potential

$V(x) = \frac{1}{2} m\omega^2 (|x| - d)^2 \quad (|x| > d).$

Assuming that the particle moves in a straight line, sketch the phase curves and show that

$E = (\sqrt{I\omega + k^2} - k)^2 \quad (k = \omega d \sqrt{2m}/\pi),$

where I is the action. Show that as the energy approaches zero the period tends to infinity, whilst at large energies it tends to $2\pi/\omega$.

(8) By expressing the generating function $S_1(\theta, q)$ for the linear oscillator, equation (7.42), as a function of (θ, I) show directly that it is periodic in θ. Similarly show directly that $S_2(I, q)$, equation (7.41), increases by $2\pi I$ as θ increases through 2π.

(9) Consider the motion of a particle of mass m in the potential

$V(q) = \frac{1}{2} m\lambda^2 (x + a)^2 \quad (x \leqslant 0)$

$\quad\quad\quad = \frac{1}{2} m\lambda^2 (x - a)^2 \quad (x \geqslant 0).$

Sketch the potential and consequent phase diagram, paying particular attention to those phase curves in the neighbourhood of the origin. Show that phase space comprises three invariant regions in each of which different action variables can be defined. In each case find the action as a function of the energy.

(10) A particle of mass m moves in the potential

$$V = A \left[a^2 - (x-a)^2 \right] \qquad (0 \leqslant x \leqslant 2a, \, A > 0)$$

$$= 0 \qquad\qquad\qquad (x \geqslant 2a)$$

and bounces elastically off a wall at $x = 0$. Sketch the potential and consequent phase curves showing clearly all regions of libration and unbounded motion.

Show that the action is given by

$$I = \frac{a^2 \sqrt{2mA}}{2\pi} \left[\epsilon - \frac{1}{2} (1 - \epsilon^2) \ln \left(\frac{1+\epsilon}{1-\epsilon} \right) \right] \quad (\epsilon < 1),$$

where the energy is $E = A a^2 \epsilon^2$. Show that $2\pi \mathrm{d}I/\mathrm{d}E$ is the period of the librating motion and hence that, as $\epsilon \to 1$ the period approaches infinity as $-\ln(1 - \epsilon)$. (Note that this type of logarithmic behaviour is typical of periodic motion in the neighbourhood of a bounded separatrix.)

(11) (a) Sketch the phase diagram for the motion of a particle in the potential

$$V(q) = \tfrac{1}{2} U (1 - \tanh \alpha q) \quad (U > 0, \alpha > 0).$$

Discuss the form of the phase curves as $\alpha \to \infty$. Hence sketch the phase curves for the potential

$$V(q) = U \quad (q < 0)$$

$$= 0 \quad (q \geqslant 0)$$

and describe the motion for particles approaching from the right with energy $E > U$ and $E < U$.

(b) Sketch the phase curves and describe the motion of a particle of mass m moving in the periodic potential

$$V(\psi) \quad = 0 \qquad (0 \leqslant \psi \leqslant \alpha \leqslant \pi)$$

$$= U \qquad (\alpha < \psi \leqslant \pi),$$

$$V(-\psi) = V(\psi),$$

where U is a positive constant. Find the energy and frequency of the motion as a function of the action in the two cases $E > U$ and $E < U$.

(12) A particle of mass m slides smoothly on a plane inclined at an angle α to the horizontal. The plane is attached to a vertical wall off which the particle bounces elastically. Sketch the phase curves, and show that the frequency of the motion is

$$\omega = \frac{2}{3}\left(\frac{3\pi mg}{2\cos\alpha}\right)^{\frac{2}{3}}\left(\frac{1}{2mI}\right)^{\frac{1}{3}},$$

where I is the action. Find the angle variable θ as a function of q, the distance of the particle along the plane measured from the wall.

(13) Sketch the phase curves and, where they exist, find the angle–action variables for the systems with Hamiltonians

(a) $p^2/2m + A(e^{-2\alpha q} - 2e^{-\alpha q})$,

(b) $p^2/2m - A\operatorname{sech}^2\alpha q$,

where A and α are positive constants.

(14) (a) The action variable defined in equation (7.14) does not exist unless the motion is periodic. However the equivalent integral for unbound motion does have a physical meaning. Consider a particle moving along the real axis under the influence of the potential $V(q)$. Let the total energy E be sufficiently large for there to be no turning points, $E > \max(V(q))$, and let the particle have positive velocity. Define the action $A(E)$ to be the difference between the action of the motion in the potential V and that of the free motion:

$$A(E) = \sqrt{2m}\int_{-\infty}^{\infty} dq(\sqrt{E - V(q)} - \sqrt{E}).$$

Suppose that the potential is

$$V(q) = U(1 - q^2/a^2) \quad (|q| \leqslant a)$$
$$= 0 \quad (|q| > a).$$

Show explicitly that $\partial A/\partial E$ is the time delay caused by the potential.

(b) Show that $\partial A/\partial E$ is the time delay for any potential $V(q)$ for which $V(q)$ tends to zero sufficiently rapidly with increasing $|q|$ for A to exist. Use the generating function $W(q_2, q_1; E)$ of exercise 6.18.

(15) Consider the time-dependent Hamiltonian

$$H = \frac{1}{2}(p^2 + \omega^2 q^2) - \frac{A}{2\omega}(p\cos\Omega t + \omega q\sin\Omega t),$$

where A, ω and Ω are positive constants, which describes a linear oscillator perturbed by a time- and velocity-dependent potential. Show that in the (θ, I) representation, where (θ, I) are the angle–action variables of the linear oscillator, example (7.5), the Hamiltonian is

$$K(\theta, I, t) = \omega I - A \left(\frac{I}{2\omega} \right)^{\frac{1}{2}} \cos (\theta - \Omega t).$$

By making the further time-dependent canonical transformation

$$\phi = \theta - \Omega t, J = I,$$

show that this is reduced to the time-independent Hamiltonian

$$K_1(\phi, J) = (\omega - \Omega) J - A \left(\frac{2J}{\omega} \right)^{\frac{1}{2}} \cos \phi.$$

Sketch the phase curves in the (ϕ, J) representation when $\omega \neq \Omega$ and the resonant case $\omega = \Omega$; in the latter case show that the amplitude of the motion is unbounded as $t \to \infty$.

Consider the non-linear Hamiltonian

$$K_2(\theta, I, t) = K (\theta, I, t) + \alpha^2 I^2,$$

where α is a constant. Show that for all real non-zero α and all Ω the amplitude of motion is bounded for all time.

In the case $\Omega = \omega$ and small α put approximate limits on the variation of $I(t)$.

8 PERTURBATION THEORY

8.1 Introduction

Real problems rarely have Hamiltonians whose equations of motion have simple solutions depending upon elementary functions, so approximation methods are very important. One such approximation applies to systems with Hamiltonians of the form

$$H(q,p) = H_0(q,p) + \epsilon H_1(q,p) \quad (|\epsilon| \ll 1), \tag{8.1}$$

where the $H_0(q,p)$ system is soluble, but for $\epsilon \neq 0$ the $H(q,p)$ system is not. This method is called *perturbation theory*. The Hamiltonians $H(q,p)$ and $H_0(q,p)$ are named the *perturbed* and *unperturbed* Hamiltonian respectively, and $\epsilon H_1(q,p)$ is named the *perturbation*. We consider the perturbation as a function of ϵ.

The most famous system of this kind is the Solar System. A given planet is predominantly affected by the Sun, but all the other planets have small effects. For example, the motion of Venus around the Sun is perturbed most by Jupiter, whose mean force on Venus is less than 2×10^{-5} times that of the Sun; the next strongest perturbation is due to the Earth whose mean force is less than 4×10^{-6} times that of the Sun. Thus, as a first approximation to the motion of Venus, we may neglect all influences other than that of the Sun. This gives a Hamiltonian whose equations of motion have simple solutions. The effect of Jupiter and the other planets may then be considered as a small perturbation to this motion. However, this is a complicated system of many degrees of freedom and our examples are all much simpler.

The central idea of perturbation theory is to expand the solution as a power series in ϵ. This is similar in spirit to the Taylor expansion of a function $f(x)$ about x:

$$f(x + \epsilon) = f(x) + \epsilon f'(x) + \frac{\epsilon^2}{2} f''(x) + \ldots \tag{8.2}$$

As with the Taylor series, a perturbation series can diverge if ϵ is too large, or indeed can diverge for all ϵ. Thus care is necessary in its use.

For Hamiltonian systems both conjugate variables need to be expanded in such a power series; consequently the resulting algebra is quite complicated and can obscure the essential ideas. In order to illustrate these ideas without getting involved in too much algebra we first consider a simple example that illustrates the theory and some of the difficulties that arise.

Example 8.1

Find a solution of a first-order system, whose motion satisfies the equation

$$\frac{dx}{dt} = x + \epsilon x^2$$

$$x(0) = A \quad (0 < \epsilon \ll 1), \tag{8.3}$$

as a power series in ϵ, correct to second order in ϵ.

We write the solution in the form

$$x(t) = x^{(0)}(t) + \epsilon x^{(1)}(t) + \epsilon^2 x^{(2)}(t) + O(\epsilon^3), \tag{8.4}$$

where each $x^{(k)}(t)$ is independent of ϵ. On substituting this expression into the equation of motion we obtain

$$\dot{x}^{(0)} + \epsilon \dot{x}^{(1)} + \epsilon^2 \dot{x}^{(2)} = x^{(0)} + \epsilon x^{(1)} + \epsilon^2 x^{(2)} + \epsilon(x^{(0)} + \epsilon x^{(1)})^2 + O(\epsilon^3). \tag{8.5}$$

Equating powers of ϵ in this Taylor expansion we obtain

$$\dot{x}^{(0)} = x^{(0)}, \tag{8.6a}$$

$$\dot{x}^{(1)} = x^{(1)} + (x^{(0)})^2, \tag{8.6b}$$

$$\dot{x}^{(2)} = x^{(2)} + 2x^{(0)}x^{(1)}. \tag{8.6c}$$

We seek a solution satisfying $x(0) = A$ for all ϵ, so equating coefficients of ϵ in equation (8.4) at time $t = 0$ gives $x^{(0)}(0) = A$, $x^{(k)}(0) = 0$ $(k \geqslant 1)$. Thus for $x^{(0)}$ we obtain

$$x^{(0)}(t) = A e^t \tag{8.7}$$

as the solution of the unperturbed equation. Substituting this into (8.6b) gives

$$\dot{x}^{(1)} = x^{(1)} + A^2 e^{2t}, \quad x^{(1)}(0) = 0, \tag{8.8}$$

with the solution

$$x^{(1)}(t) = A^2 e^t (e^t - 1). \tag{8.9}$$

Substituting $x^{(0)}(t)$ and $x^{(1)}(t)$ into (8.6c) gives

$$\dot{x}^{(2)} = x^{(2)} + 2A^3 e^{2t}(e^t - 1), \quad x^{(2)}(0) = 0, \tag{8.10}$$

so that

$$x^{(2)}(t) = A^3 e^t (e^t - 1)^2. \tag{8.11}$$

In general each term in the perturbation expansion is obtained from the previous terms.

To second order in ϵ the solution is

$$x(t) = Ae^t \left[1 + \epsilon A(e^t - 1) + \epsilon^2 A^2 (e^t - 1)^2 \right] + O(\epsilon^3). \tag{8.12}$$

We see that each term in the square brackets is $\epsilon A(e^t - 1)$ times the preceding term. Thus, for a given accuracy, the first-order perturbation expansion, $x^{(0)} + \epsilon x^{(1)}$, is useful for a shorter time than the second-order expansion, equation (8.12). Conversely, if we require a given accuracy we must work to higher orders in the perturbation expansion, the larger the time. These are common properties of perturbation expansions, and are illustrated in figure 8.1.

In this particular example the equation of motion (8.3) can be integrated directly to give

$$x(t) = \frac{Ae^t}{1 - \epsilon A(e^t - 1)}, \tag{8.13}$$

which is seen to give (8.12) when expanded in powers of ϵ. We also notice that the power series expansion becomes invalid at a critical time given by

$$\epsilon A(e^{t_c} - 1) = 1 \quad \text{or} \quad t_c = \ln\left(\frac{1 + A\epsilon}{A\epsilon}\right). \tag{8.14}$$

Thus, the perturbation series expansion is valid only for times $t < t_c$ no matter how many orders of perturbation are calculated. The reason for this is that the unperturbed equation,

$$\dot{x} = x, \quad x(0) = A, \tag{8.15}$$

describes motion which does not terminate, whilst the motion of the perturbed system terminates, so that the 'small perturbation' ϵx^2 alters the type of motion. Nevertheless the perturbation expansion does give a useful approximation to the solution for small times.

In figure 8.1 we show the exact solution divided by the unperturbed solution, named the exact quotient,

$$f(t) = \frac{x(t)}{Ae^t} = \frac{1}{1 - \epsilon A(e^t - 1)} \tag{8.16}$$

and the approximate quotient according to Nth-order perturbation theory,

$$f^{(N)}(t) = \sum_{n=0}^{N} [\epsilon A(e^t - 1)]^n. \tag{8.17}$$

for $N = 0, 1, 2, 4, 10$. In the illustration $A = 1$ and $\epsilon = 0.1$ so that $t_c = 2.4$.

Fig. 8.1 Graph comparing the quotient $f(t)$, equation (8.16), shown by the solid line with the approximate quotients $f^{(N)}(t)$, equation (8.17), for $N = 0, 1, 2, 4, 10$, shown by the dashed lines.

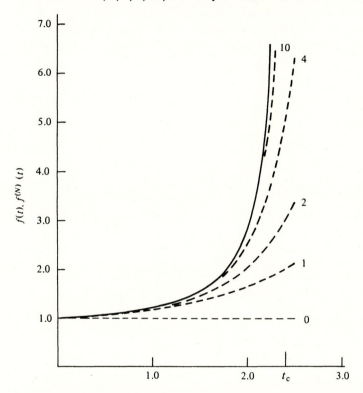

This example is very simple, but it illustrates the techniques and some of the difficulties of perturbation expansion. The reader is advised to try exercises 8.1 to 8.7 before proceeding to the next section where the method is used extensively for the more complicated Hamiltonian systems.

8.2   First-order perturbation theory for conservative Hamiltonian systems

Suppose that the Hamiltonian is the sum of two terms:

$$H(q, p) = H_0(q, p) + \epsilon H_1(q, p), \qquad (8.18)$$

where Hamilton's equations for H_0 have known solutions and the perturbation ϵH_1 is small.

In perturbation theory we express the solution of Hamilton's equations as a power series in ϵ, as in example 8.1. Since this is an expansion about the unperturbed motion, the perturbed motion must be of the same type, that is both are rotations, or both librations.

Consider, for example, the Hamiltonian of the vertical pendulum, equations (4.43) and (5.30)

$$H(\psi, p) = \tfrac{1}{2} p^2 - \alpha^2 \cos \psi, \tag{8.19}$$

which is discussed in section 4.7 and whose phase curves are shown in figure 8.2. The phase space is divided by the separatrix s into regions of rotation and regions of libration.

For fast rotations, where

$$\tfrac{1}{2} p^2 \gg \alpha^2, \tag{8.20}$$

the potential energy term is relatively small and we choose

$$H_0 = \tfrac{1}{2} p^2, \quad \epsilon H_1 = -\alpha^2 \cos \psi. \tag{8.21}$$

On the other hand, to investigate libration near the equilibrium point at 0, we write the Hamiltonian as

$$H(\psi, p) = \frac{1}{2} (p^2 + \alpha^2 \psi^2) - \alpha^2 - \alpha^2 (\cos \psi - 1 + \tfrac{1}{2} \psi^2), \tag{8.22}$$

Fig. 8.2 The phase diagram of the vertical pendulum with the Hamiltonian of equation (8.19). The separatrix s divides the region of libration, the closed curves centred upon 0, from the regions of rotation.

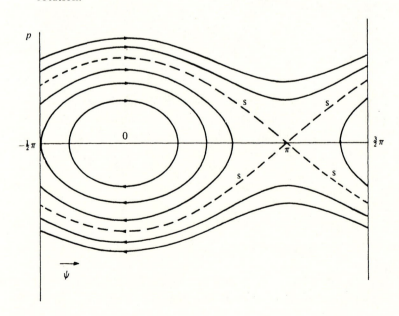

so that, ignoring the constant $-\alpha^2$, we have different unperturbed and perturbed Hamiltonians,

$$\bar{H}_0 = \tfrac{1}{2}\,(p^2 + \alpha^2\,\psi^2) \quad \text{and} \quad \epsilon\bar{H}_1 = -\alpha^2\,(\cos\psi + \tfrac{1}{2}\,\psi^2). \tag{8.23}$$

The unperturbed motion resembles a linear oscillator, the more usual 'pendulum motion'. Hamilton's equations for H_0 and $\bar{H}_0$ have simple solutions; see examples 8.2 and 8.3.

We now continue with the general analysis to first order only in ϵ. It applies to both rotation and libration. We assume that there is a region of phase space in which the phase curves of H and H_0 may be continuously deformed into each other. We also assume that the old angle–action variables (ϕ, J) of H_0 and the new variables (θ, I) of H may be related to each other by a power series expansion in ϵ:

$$\phi = \phi^{(0)}(\theta, I) + \epsilon\,\phi^{(1)}(\theta, I) + O(\epsilon^2) \tag{8.24a}$$

$$J = J^{(0)}(\theta, I) + \epsilon\,J^{(1)}(\theta, I) + O(\epsilon^2), \tag{8.24b}$$

where each $(\phi^{(k)}, J^{(k)})$ is independent of ϵ.

We chose to express the old angle–action variables (ϕ, J) in terms of the new (θ, I) because, for the perturbed motion, I = constant and $\theta = \Omega t$ + constant, where Ω is the perturbed frequency. Once the expansions (8.24) and this frequency have been found the time dependence of (ϕ, J) is known, so the phase curve is defined parametrically in terms of θ or t.

For the unperturbed system, $\epsilon = 0$, $\theta = \phi$ and $I = J$, and consequently

$$\phi^{(0)}(\theta, I) = \theta \quad J^{(0)}(\theta, I) = I. \tag{8.25}$$

The problem now is to find $(\phi^{(1)}, J^{(1)})$ and the Hamiltonian in the (θ, I) representation.

The phase curves of the perturbed and unperturbed motions in each representation are sketched in the diagram below, where the perturbed motion is shown by a solid line and the unperturbed motion by a broken one.

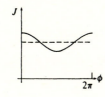

When θ increases by 2π, so does ϕ, so that $\phi^{(1)}(\theta, I)$, like $J^{(1)}(\theta, I)$, must be periodic in θ with period 2π. Since angle variables are defined to within an additive constant, it is convenient to choose $\phi^{(1)}$ to have zero mean value when averaged over θ.

Since the transformation $(\phi, J) \to (\theta, I)$ is canonical, it follows from the theory of chapter 6 that it is area-preserving, so that the Jacobian of this transformation is unity,

$$\frac{\partial(\phi, J)}{\partial(\theta, I)} = 1. \tag{8.26}$$

Also, the area under the perturbed phase curve is the same in each representation. In the (θ, I) representation I is constant along each perturbed phase curve, but in the (ϕ, J) representation J is a periodic function of ϕ, therefore

$$I = \frac{1}{2\pi} \int_0^{2\pi} d\theta \, I = \frac{1}{2\pi} \int_0^{2\pi} d\phi J(\phi) \,. \tag{8.27}$$

Using these periodicity and area-preserving relations we can find $(\phi^{(1)}, I^{(1)})$ and the Hamiltonian in the (θ, I) representation. By definition, since I is an action variable of the perturbed system, this Hamiltonian is independent of θ. It can be expanded in powers of ϵ as

$$K(I) = K_0(I) + \epsilon K_1(I) + O(\epsilon^2) \,. \tag{8.28}$$

But in the (ϕ, J) representation the Hamiltonian is

$$H(\phi, J) = H_0(J) + \epsilon H_1(\phi, J) \tag{8.29}$$

and the new Hamiltonian (8.28) may be obtained by expressing (ϕ, J) as functions of (θ, I), equation (8.24), and substituting into (8.29):

$$K(I) = H_0(I + \epsilon J^{(1)}) + \epsilon H_1(\theta, I) + O(\epsilon^2)$$

$$= H_0(I) + \epsilon \left[J^{(1)} \frac{\partial H_0}{\partial I} + H_1(\theta, I) \right] + O(\epsilon^2) \,. \tag{8.30}$$

By equating the coefficients of ϵ of equations (8.30) and (8.28) we find

$$K_0(I) = H_0(I), \tag{8.31}$$

$$K_1(I) = J^{(1)}(\theta, I) \frac{\partial H_0}{\partial I}(I) + H_1(\theta, I) \,. \tag{8.32}$$

Equation (8.31) gives $K_0(I)$ directly; $K_1(I)$ and $J^{(1)}(\theta, I)$ are found using the periodicity and area-preserving properties. On substituting (8.24) into the right-hand side of (8.27) we find

$$I = \frac{1}{2\pi} \int_0^{2\pi} d\theta \left(1 + \epsilon \frac{\partial \phi^{(1)}}{\partial \theta} \right) (I + \epsilon J^{(1)}) + O(\epsilon^2)$$

$$= I + \frac{\epsilon}{2\pi} \int_0^{2\pi} d\theta \, J^{(1)} + O(\epsilon^2), \tag{8.33}$$

the last line following because $\phi^{(1)}$ is periodic. Thus

$$\int_0^{2\pi} d\theta \, J^{(1)}(\theta, I) = 0 \tag{8.34}$$

and, on taking the average of (8.32) over θ, we obtain

$$K_1(I) = \frac{1}{2\pi} \int_0^{2\pi} d\theta \, H_1(\theta, I). \tag{8.35}$$

This shows that *the first-order correction to the Hamiltonian is the mean of the perturbation, ϵH_1, taken over the unperturbed motion.* Since K_1 is now known we can find $J^{(1)}(\theta, I)$ directly from (8.32):

$$J^{(1)} = \frac{K_1(I) - H_1(\theta, I)}{\omega_0(I)} \quad (\omega_0(I) \neq 0) \tag{8.36}$$

where $\omega_0 = \partial H_0 / \partial I$ is the unperturbed frequency, which must be non-zero for this theory to be valid.

If ω_0 is small the effect of the perturbation can be quite large. Since the nth-order perturbation has terms proportional to $(\omega_0)^{-n}$ (see for example exercise 8.15), the convergence of the perturbation expansion is doubtful. Generally ω_0 is small near a separatrix dividing phase space into regions containing different types of motion. This problem becomes much more serious for systems of N degrees of freedom ($N \geqslant 2$) because then there are N fundamental frequencies. These and all their integer linear combinations occur in the denominator of the expressions equivalent to (8.36), so that, even if all the frequencies are large, a particular combination may be small. Thus the convergence of perturbation expansions for systems of more than one degree of freedom is a very subtle problem.

Having found $J^{(1)}$ we can use equation (8.26) to find $\phi^{(1)}(\theta, I)$. On substituting (8.24) into (8.26) and equating powers of ϵ we find that

$$\frac{\partial \phi^{(1)}}{\partial \theta} = -\frac{\partial J^{(1)}}{\partial I}. \tag{8.37}$$

This may be integrated directly to give $\phi^{(1)}$, remembering that the constant of integration is chosen to make the mean of $\phi^{(1)}$ equal to zero.

Example 8.2

Consider the motion of a vertical pendulum which is rotating fast enough for the gravitational forces to be considered as a small perturbation. As discussed at the beginning of this section, the unperturbed Hamiltonian and the perturbation are

$$H_0 = \tfrac{1}{2} p^2, \quad H_1 = -\alpha^2 \cos \psi, \tag{8.38}$$

where we have put $\epsilon = 1$, and we treat α as the small parameter. The angle–action variables for the unperturbed system are simply

$$\phi = \psi, \quad J = p. \tag{8.39}$$

Then, from equations (8.31) and (8.35), the first two terms of the Hamiltonian in the (θ, I) representation are

$$K_0(I) = \tfrac{1}{2} I^2 \tag{8.40}$$

$$K_1(I) = -\frac{\alpha^2}{2\pi} \int_0^{2\pi} d\theta \, \cos \theta = 0. \tag{8.41}$$

Also from (8.36) and (8.37)

$$J^{(1)} = \frac{\alpha^2 \cos \theta}{I} \tag{8.42}$$

and

$$\frac{\partial \phi^{(1)}}{\partial \theta} = \frac{\alpha^2 \cos \theta}{I^2}, \tag{8.43}$$

so that

$$\phi^{(1)} = \frac{\alpha^2 \sin \theta}{I^2}. \tag{8.44}$$

Thus the perturbed motion is

$$\phi = \psi = \theta + \frac{\alpha^2 \sin \theta}{I^2} + O(\alpha^4) \tag{8.45}$$

$$J = p = I + \frac{\alpha^2 \cos \theta}{I} + O(\alpha^4) \tag{8.46}$$

and the Hamiltonian is

$$K(I) = \tfrac{1}{2} I^2 + O(\alpha^4). \tag{8.47}$$

The angle variable θ varies linearly with time,

$$\dot{\theta} = \frac{\partial K}{\partial I} = I, \quad \theta = It + \delta, \tag{8.48}$$

which gives the time dependence of the physical angle ψ and the angular momentum p:

$$\psi = It + \delta + \frac{\alpha^2 \sin (It + \delta)}{I^2} + O(\alpha^4), \tag{8.49}$$

$$p = I + \frac{\alpha^2 \cos (It + \delta)}{I} + O(\alpha^4). \tag{8.50}$$

Because $K_1(I) = 0$, to this order of approximation the frequency of the perturbed and unperturbed motions having the same action are the same. This is always true for first-order perturbations whose mean is zero.

Example 8.3
Consider the motion of a vertical pendulum executing oscillations about the downward vertical. As discussed at the beginning of this section, the unperturbed Hamiltonian is

$$H_0(\psi, p) = \tfrac{1}{2} (p^2 + \alpha^2 \psi^2), \tag{8.51}$$

and, for $\epsilon = 1$, the perturbation is

$$H_1(\psi, p) = -\alpha^2 (\cos \psi - 1 + \tfrac{1}{2} \psi^2). \tag{8.52}$$

In this example α need not be small; the perturbation is small because ψ is small. We may expand $\cos \psi$ in a Taylor series to obtain

$$H_1 = -\tfrac{1}{24} \alpha^2 \psi^4 + O(\psi^6). \tag{8.53}$$

Since we are working to first order, the perturbation may be written

$$H_1 = -\tfrac{1}{24} \alpha^2 \psi^4. \tag{8.54}$$

In this problem there is no naturally occurring small parameter ϵ and in these circumstances it is usually useful to introduce an ϵ which we set equal to unity at the end of the problem. Thus we write the Hamiltonian as

$$H = \tfrac{1}{2} (p^2 + \alpha^2 \psi^2) - \tfrac{1}{24} \epsilon \alpha^2 \psi^4 + O(\epsilon^2). \tag{8.55}$$

By example 7.5, the angle-action variables for H_0 are (ϕ, J), where

$$\psi = \left(\frac{2J}{\alpha} \right)^{\frac{1}{2}} \sin \phi, \quad p = (2\alpha J)^{\frac{1}{2}} \cos \phi, \tag{8.56}$$

and, to first order in ϵ, the Hamiltonian in this representation is

$$H(\phi, J) = \alpha J - \tfrac{1}{6} \epsilon J^2 \sin^4 \phi, \tag{8.57}$$

where the first term is H_0 and the second is ϵH_1. Then, from equations (8.31) and (8.35), we find that the first two terms of the Hamiltonian in (θ, I) representation are given by

$$K_0(I) = \alpha I \tag{8.58}$$

$$K_1(I) = -\tfrac{1}{6} I^2 \int_0^{2\pi} d\theta \, \sin^4\theta = -\tfrac{1}{16} I^2 . \tag{8.59}$$

To first order in ϵ the Hamiltonian K and frequency Ω of the perturbed motion are

$$K(I) = \alpha I - \tfrac{1}{16} \epsilon I^2 + O(\epsilon^2), \tag{8.60}$$

$$\Omega = \partial K / \partial I = \alpha - \tfrac{1}{8} \epsilon I + O(\epsilon^2). \tag{8.61}$$

From equations (8.36) and (8.37), $J^{(1)}$ and $\phi^{(1)}$ are

$$J^{(1)} = -\frac{I^2}{\alpha} (\tfrac{1}{16} - \tfrac{1}{6} \sin^4\theta)$$

$$= -\frac{I^2}{12\alpha} (\cos 2\theta - \tfrac{1}{4} \cos 4\theta) \tag{8.62}$$

$$\phi^{(1)} = \frac{I}{12\alpha} (\sin 2\theta - \tfrac{1}{8} \sin 4\theta), \tag{8.63}$$

so that

$$\phi = \theta + \frac{\epsilon I}{12\alpha} [\sin 2\theta - \tfrac{1}{8} \sin 4\theta] + O(\epsilon^2) \tag{8.64}$$

$$J = I - \frac{\epsilon I^2}{12\alpha} [\cos 2\theta - \tfrac{1}{4} \cos 4\theta] + O(\epsilon^2), \tag{8.65}$$

where

$$\theta = \Omega t + \delta, \tag{8.66}$$

the angular frequency Ω being given by (8.61).

The motion of the original variables (ψ, p) may be obtained by substituting (8.64) and (8.65) into (8.56):

$$\psi = \left(\frac{2I}{\alpha}\right)^{\frac{1}{2}} \left[1 - \frac{\epsilon I}{12\alpha} (\cos 2\theta - \tfrac{1}{4} \cos 4\theta) \right]^{\frac{1}{2}}$$

$$\times \quad \sin \left[\theta + \frac{\epsilon I}{12\alpha} (\sin 2\theta - \tfrac{1}{8} \sin 4\theta) \right] + O(\epsilon^2). \tag{8.67}$$

This may be simplified by expanding in ϵ and retaining only terms linear in ϵ:

$$\psi = \left(\frac{2I}{\alpha}\right)^{\frac{1}{2}} \left[\sin \theta + \frac{\epsilon I}{48\alpha} (3 \sin \theta + \tfrac{1}{2} \sin 3\theta) \right]. \tag{8.68}$$

Similarly we obtain

$$p = (2\alpha I)^{\frac{1}{2}} \left[\cos\theta - \frac{\epsilon I}{32\alpha} \left(2\cos\theta - 3\cos 3\theta\right) \right]. \tag{8.69}$$

To this order the perturbed and unperturbed phase curves are approximately a distance ϵI apart. This does not mean that the perturbed and unperturbed values of $\psi(t)$ and $p(t)$ differ by this quantity for all t, because the frequencies with which each phase curve is traversed are different. In a time T the unperturbed phase point rotates through an angle $\phi = \alpha T$, whilst the perturbed phase point rotates through an angle $\theta = \Omega T$, so that, from (8.61),

$$\phi - \theta = \tfrac{1}{8}\,\epsilon I T.$$

Thus, after a time $8\pi/\epsilon I$ the perturbed and unperturbed motions are $180°$ out of phase.

For $\epsilon = 1$ the frequency Ω of the perturbed motion, given by equation (8.61), is zero when $I = 8\alpha$, which, from the Hamiltonian (8.60), is at an energy $4\alpha^2$. This is the first-order approximation to the separatrix energy of (8.55) which is $\tfrac{3}{2}\alpha^2$. This in turn is an approximation to the separatrix energy of the vertical pendulum, which is α^2, as shown in section 4.6.

The first-order approximation to the separatrix energy is very poor. In general, perturbation theory gives a poor approximation to any phase curve near a separatrix, because the frequency of these motions is always small and from equation (8.36) we see that this always appears in the denominator of the perturbed solution.

Exercises for chapter 8

(1) Find a perturbation expansion correct to second order in ϵ for the solution of the algebraic equation

$x = 1 + \epsilon x^2$,

where ϵ is a small parameter. Compare your expansion with the exact solution. Note that a perturbation can introduce a new solution.

(2) Show that a solution of the algebraic equation

$x = a + \epsilon x^\alpha \quad (x > 0, a > 0, \alpha > 1)$

which is correct to third order in the small parameter ϵ is

$$x = a + \epsilon a^\alpha + \epsilon^2 \alpha\, a^{2\alpha - 1} + \epsilon^3\, \frac{\alpha}{2}\,(3\alpha - 1)a^{3\alpha - 2} + O(\epsilon^4).$$

(3) In the example above treat a as the small perturbation parameter and show that there is another root at

$$x \simeq \left(\frac{1}{\epsilon}\right)^{1/(\alpha-1)} - \left(\frac{a}{\alpha - 1}\right).$$

(4) Find the roots of the equation

$$\sin (x^2 + \epsilon \sin x) = 0 \quad (|\epsilon| \ll 1),$$

correct to first order in ϵ.

(5) Show that a solution of the algebraic equation

$$x = a + \epsilon f(x),$$

which is correct to third order in the small parameter ϵ is

$$x = a + \epsilon f(a) + \epsilon^2 f(a) f'(a) + \epsilon^3 \left[f(a) f'(a)^2 + \tfrac{1}{2} f(a)^2 f''(a) \right],$$

where $f(x)$ is any sufficiently well-behaved function.

(6) Find a perturbation series solution, correct to second order in the small parameter ϵ, for the first-order differential equation

$$\frac{dx}{dt} = x + \epsilon x^\alpha \quad (\alpha \neq 1, x(0) = A > 0).$$

Compare your series with the exact solution and discuss the convergence of the perturbation expansion in the cases $0 < \alpha < 1$ and $\alpha > 1$.

(7) Find a perturbation series solution correct to first order in ϵ for each of the differential equations

(a) $\dfrac{dx}{dt} = x^{\frac{1}{2}} + \epsilon x^{\frac{1}{4}} \quad (x(0) = A > 0)$

(b) $\dfrac{dx}{dt} = \sin x + \epsilon x \quad (x(0) = A, 0 < A < \pi).$

(8) Use perturbation theory to obtain the first-order corrections to the motion of the linear oscillator with Hamiltonian

$$H_0(q, p) = \tfrac{1}{2} (p^2 + q^2)$$

perturbed by $\tfrac{1}{2} \epsilon q^2$, that is, determine $E(I)$ and $q(t)$ to first order in ϵ. Compare your results with the exact solution.

(9) Use perturbation theory to obtain the first-order correction to the motion of the free rotor with Hamiltonian

$$H_0(p) = p^2 / 2G,$$

when perturbed by the potentials

(a) $V(\psi) = \alpha \sin^3 \psi,$

(b) $V(\psi) = \alpha \sin^4 \psi.$

(10) Find the first-order corrections to the motion of the linear oscillator with Hamiltonian

$H(q,p) = p^2/2m + \frac{1}{2} m\omega^2 q^2$

perturbed by

(a) the quintic potential $\epsilon H_1 = \epsilon q^5$,

(b) the velocity dependent potential $\epsilon H_1 = \epsilon(q^4 + \alpha q p^2)$, where α is a constant.

(11) A particle of mass m oscillates about the minimum of the potential

$V = A(q^{-2n} - 2q^{-n})$,

A being a positive constant and n an integer. Show that an approximate Hamiltonian for the motion is given by

$H = p^2/2m + An^2x^2 - An^2(n+1)x^3 \quad (x = q - 1)$

and find an approximation to the motion, correct to first order. Would a second-order treatment of H be of any value?

(12) A particle of mass m slides under the influence of gravity on a smooth rigid wire in the shape $z = \frac{1}{2}\alpha^2 x^2$, where the z-axis is vertically upwards. Find the Hamiltonian for the system and show that an approximation to this Hamiltonian, correct to order $(px)^2$, p being the momentum, is

$H = \dfrac{p^2}{2m} + \dfrac{mg\alpha^2 x^2}{2} - \dfrac{\alpha^4}{2m} x^2 p^2.$

Hence find an approximation to the motion which is correct to this order.

(13) Find the first-order change in the frequency and motion of a particle of negative total energy in the Morse potential

$V(q) = A(e^{-2\alpha q} - 2e^{-\alpha q}) \quad (A > 0, \alpha > 0)$,

when the small potential

$V_1(q) = -\epsilon e^{\alpha q} \quad (0 < \epsilon \ll A)$

is added to it.

(14) A particle of mass m is held by two springs as shown in the diagram. Each spring has natural length l and exerts a restoring force $k(s-l)$, where k is a positive constant, when its length is $s(>l)$. In the equilibrium position, each spring needs to be stretched a length d to join. If the particle is displaced a distance x, as shown in the diagram,

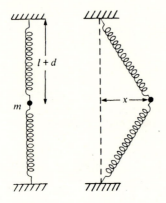

show that if the effect of gravity is negligible the potential energy of the particle is

$$V(x) = k \left[\frac{dx^2}{l+d} + \frac{lx^4}{4(l+d)^3} \right] + O(x^6).$$

In the case $d = 0$ determine how the frequency of the oscillations depends upon the amplitude. When $d \neq 0$ determine an approximation to the frequency and the motion $x(t)$ commensurate with the accuracy of the potential.

(15) Show that the second-order terms in the expansion of the old angle-action variables in terms of the new, $(\phi^{(2)}, J^{(2)})$ equation (8.24), and the Hamiltonian, $K_2(I)$ equation (8.28), satisfy the equations

$$\frac{\partial J^{(2)}}{\partial I} + \frac{\partial \phi^{(2)}}{\partial \theta} = \frac{-\partial(\phi^{(1)}, J^{(1)})}{\partial(\theta, I)},$$

$$\int_0^{2\pi} d\theta \left[J^{(2)} + J^{(1)} \frac{\partial \phi^{(1)}}{\partial \theta} \right] = 0,$$

$$K_2 = J^{(2)} \frac{\partial H_0}{\partial I} + \frac{1}{2} [J^{(1)}]^2 \frac{\partial^2 H_0}{\partial I^2} + \phi^{(1)} \frac{\partial H_1}{\partial \theta} + J^{(1)} \frac{\partial H_1}{\partial I}.$$

Hence show that the second-order correction to the Hamiltonian is

$$K_2(I) = \frac{1}{4\pi} \frac{\partial}{\partial I} \int_0^{2\pi} d\theta \frac{K_1^2 - H_1^2}{\omega_0} \qquad \left(\omega_0 = \frac{\partial H_0}{\partial I} \right).$$

(16) Use the result of the previous exercise to show that the energy and frequency of rotation of the rapidly rotating vertical pendulum considered in example 8.2 are

$$K(I) = \tfrac{1}{2} I^2 + \alpha^4/4I^2 + O(\alpha^6),$$

$$\Omega(I) = I(1 - \alpha^4/2I^4) + O(\alpha^6).$$

(17) Show that the Hamiltonian

$$H = \tfrac{1}{2} p^2 + p \sin q$$

has an elliptic fixed point at $q = \tfrac{1}{2}\pi$, $p = -1$ in whose neighbourhood

$$H = -\tfrac{1}{2} + \tfrac{1}{2}(P^2 + Q^2) - \tfrac{1}{2} PQ^2 - \tfrac{1}{24} Q^4 + \text{higher terms},$$

where $q = \tfrac{1}{2}\pi + Q$, $p = P - 1$. Hence, using the result of exercise 8.15 show that the Hamiltonian and frequency can be expressed in terms of the action as follows

$$K = I - \tfrac{1}{4} I^2$$

$$\Omega = 1 - \tfrac{1}{2} I.$$

(18) Using the results of exercise 8.15 show that, for high energies, the rotational motion of the Hamiltonian

$$H = \tfrac{1}{2} p^2 + p \sin q$$

is given by the approximate equation

$$q = \theta - \frac{\sin 2\theta}{8I^2} + \dots$$

$$p = I - \sin \theta - \frac{\cos 2\theta}{4I} + \dots,$$

where to this order the angle–action variables (θ, I) satisfy Hamilton's equations for the Hamiltonian

$$K = \tfrac{1}{2} I^2.$$

(19) Suppose that a perturbed free particle of mass m has the Hamiltonian

$$H = p^2/2m + \epsilon e^{-\lambda q^2},$$

and that the energy, E, of the particle satisfies $E \gg \epsilon$. Let the unperturbed motion be

$$q = Q + (t - t_0)P/m \qquad p = P.$$

Interpret these equations as a time-dependent canonical transformation from (q, p) to (Q, P) representation and show that the new Hamiltonian is

$$K(Q, P, t - t_0) = \epsilon \exp \{-\lambda[Q + (t - t_0)P/m]^2\}.$$

Show that in this representation the equations of motion have the approximate solution

$$Q = Q_0 + \epsilon \, \partial A / \partial P_0 + O(\epsilon^2), \quad P = P_0 - \epsilon \, \partial A / \partial Q_0 + O(\epsilon^2),$$

where

$$A(Q_0, P_0, t - t_0) = \int_{t_0}^{t} dt' \, K(Q_0, P_0, t' - t_0)$$

is the integral of the perturbation over the unperturbed motion. Hence show that the motion with initial conditions $q(t_0) = Q_0 < 0$, such that $\lambda Q_0^2 \gg 1$ and moving in the positive direction $p(t_0) = P_0 > 0$, long after passing through the perturbation is

$$q(t) = Q_0 - \frac{m}{P_0^2} \sqrt{\frac{\pi}{\lambda}} + \frac{(t - t_0)P_0}{m} \qquad p(t) = P_0.$$

You will need the result

$$\int_{-\infty}^{\infty} dx \, e^{-x^2} = \sqrt{\pi}.$$

(20) Consider a system with Hamiltonian.

$$H(q, p, t) = H_0(q, p) + \epsilon H_1(q, p, t),$$

representing a conservative Hamiltonian H_0 perturbed by a time-dependent perturbation ϵH_1. Suppose that the motion of the unperturbed system is known and given by

$$q = f(Q, P, t - t_0), \quad p = g(Q, P, t - t_0),$$

where (Q, P) represents the state at $t = t_0$. Interpret these equations as a time-dependent canonical transformation and show that in the (Q, P) representation the Hamiltonian is

$$K(Q, P, t) = \epsilon H_1 \left(f(Q, P, t - t_0), g(Q, P, t - t_0), t \right).$$

Use perturbation theory to solve the equations of motion in the (Q, P) representation and hence show that the approximate motion in the (q, p) representation is

$$q = f(Q_0, P_0, t - t_0) + \epsilon \left(\frac{\partial f}{\partial Q_0} \frac{\partial A}{\partial P_0} - \frac{\partial f}{\partial P_0} \frac{\partial A}{\partial Q_0} \right)$$

$$p = g(Q_0, P_0, t - t_0) + \epsilon \left(\frac{\partial g}{\partial Q_0} \frac{\partial A}{\partial P_0} - \frac{\partial g}{\partial P_0} \frac{\partial A}{\partial Q_0} \right),$$

where

$$A(Q_0, P_0, t, t_0) = \int_{t_0}^{t} dt' H_1 \left(f(Q_0, P_0, t' - t_0), g(Q_0, P_0, t' - t_0), t' \right)$$

and $q(t_0) = Q_0$, $p(t_0) = P_0$.

(21) (a) Consider a system with Hamiltonian

$$H(q, p, t) = H_0(q, p) + \epsilon H_1(q, p, t)$$

representing a conservative Hamiltonian, H_0, perturbed by the time-dependent perturbation, ϵH_1. If H_1 goes to zero sufficiently rapidly as $|t| \to \infty$ show that the change in energy of the unperturbed system is

$$\Delta E = \epsilon \lim_{t_0 \to -\infty} \lim_{t \to \infty} \frac{\partial A}{\partial t_0} (t, t_0) + O(\epsilon^2)$$

where A is defined in exercise 8.20.

(b) Using this result, show that the first-order change in the energy of the linear oscillator

$$H_0(q, p) = p^2/2m + \tfrac{1}{2} m\omega^2 q^2,$$

with initial energy E, when perturbed by the potential

$$\epsilon V(q) = \epsilon U q e^{-\lambda |t|} \quad (\lambda > 0),$$

is

$$\Delta E = - \frac{2\lambda \epsilon U}{\lambda^2 + \omega^2} \sqrt{\frac{2E}{m}} \cos \delta,$$

where δ is the initial phase of the oscillator. You will need the result

$$\int_{0}^{\infty} dt \, e^{-\lambda t} \cos \omega t = \frac{\lambda}{\lambda^2 + \omega^2} \, .$$

(22) This example illustrates the danger of applying perturbation theory directly to Newton's equations. Show that Newton's equation of motion corresponding to the Hamiltonian of equation (8.55) is

$$\ddot{\psi} + \alpha^2 \, \psi = \tfrac{1}{6} \epsilon \, \alpha^2 \, \psi^3.$$

Write the solution to this equation in the form $\psi = \psi_0 + \epsilon \, \psi_1 + O(\epsilon^2)$ and show that

$$\ddot{\psi}_0 + \alpha^2 \, \psi_0 = 0$$

$$\ddot{\psi}_1 + \alpha^2 \, \psi_1 = \tfrac{1}{6} \, \alpha^2 \, \psi_0^3.$$

Hence show that an approximate solution having initial conditions
$\psi(0) = A$, $\dot{\psi}(0) = 0$ is

$$\psi = A\left[\cos \alpha t + \frac{A^2 \epsilon}{192} \left(12t\alpha \sin \alpha t - 3 \cos 3\alpha t + \cos \alpha t\right)\right].$$

This solution is unbounded as $t \to \infty$, whilst that found in the text, equation (8.68), is bounded. By comparing these two solutions show that the difference is a consequence of this second solution having the wrong frequency.

9 ADIABATIC AND RAPIDLY OSCILLATING CONDITIONS

9.1 Introduction

We have seen that conservative Hamiltonian systems with one degree of freedom are particularly simple to understand because the phase curves are the Hamiltonian contours. When the Hamiltonian depends explicitly upon time this simplicity is lost: a general analysis of the motion does not exist. However, the motion of related conservative systems can often be used to obtain approximate solutions. There are two extreme situations in which such approximations are useful:

(1) The motion is approximately periodic of period T and the Hamiltonian changes little during the time T. These slow changes in the Hamiltonian are named *adiabatic*.

(2) The system is acted upon by an external periodic force of period small by comparison to the time during which the unperturbed motion changes significantly.

Adiabatic theory has many applications. For example, the theory of simple pendulums of slowly varying length was once important in time-keeping and nowadays a similar theory applies to the quartz-crystal oscillators used in modern clocks. Adiabatic theory describes the motion of the planets, because this is affected by the Sun's changing mass, which changes by a factor of about 10^{-13} in a year. At the microscopic level also, adiabatic theory has its application: the molecules in a gas at low pressure in a vessel oscillate back and forth with periods from roughly 10^{-5} to 10^{-4} seconds, whereas changes in the vessel itself normally occur in times of one second or more. Similar considerations apply to charged particles in magnetic fields, with applications in astrophysics, geophysics, particle accelerators and proposed controlled thermonuclear reactors.

At the other extreme, rapidly changing forces occur in, for example, the interaction of lasers with atoms or molecules, and in machinery having rapidly rotating parts.

In this chapter we first study the effects of adiabatic change. The result obtained is very simple: the action variable is almost constant during an adiabatic change. This is stated more carefully in section 9.4, where it is proved. However, this proof is difficult and so may obscure the simplicity of the final result. For this reason, in sections 9.2 and 9.3 we give two simple examples showing explicitly that the action is an approximate constant of the motion: at a first reading the proof in section 9.4 may be omitted, but not example 9.1 at the end of that section. Finally in section 9.5 we consider the opposite extreme of a system acted upon by a rapidly oscillating force.

9.2 Elastic ball bouncing between two slowly moving planes

Consider the motion of a perfectly elastic ball of unit mass bouncing between two planes moving together with constant speed V. For simplicity we suppose that one plane is fixed and that the motion of the other plane is unaffected by the repeated collisions with the ball.

First, consider the case $V = 0$. By example 7.3, if v is the speed of the ball and x the distance between the planes, then the action is

$$I = vx/\pi \tag{9.1}$$

Now suppose that $0 < V \ll v$ and consider an approximate analysis in which the terms of order V/v are neglected. The planes move little in the time Δt between successive collisions with the moving plane. Then

$$\Delta t \simeq 2x/v. \tag{9.2}$$

In this time, by definition, the ball collides once with the moving plane and, since the relative velocity of the ball and plane is constant, the speed v' after the collision is

$$v' = v + 2V, \tag{9.3}$$

v being the speed before the collision. Thus the mean rate of change of speed is

$$\frac{dv}{dt} = \frac{v' - v}{\Delta t} = \frac{Vv}{x}. \tag{9.4}$$

The rate of change of action is approximately dI/dt, where

$$\pi \frac{dI}{dt} = (\dot{v}x + v\dot{x})$$

$$= (Vv - Vv) = 0, \tag{9.5}$$

since $\dot{x} = -V$. Thus the action is approximately constant. Note that this analysis does not assume that the planes are moving uniformly but only that $V \ll v$. The approximate constancy of a dynamical variable under slow or adiabatic changes in the Hamiltonian function is known as *adiabatic invariance*.

Now consider an exact analysis. Let v_n be the speed of the ball immediately before the nth collision with the moving plane and v_{n+1} the speed immediately afterwards. Then, as above,

$$v_{n+1} = v_n + 2V, \tag{9.6}$$

so that

$$v_n = v_0 + 2nV. \tag{9.7}$$

Let x_n be the separation of the planes at the instant of the nth collision and Δt_n the time between the $(n + 1)$th and nth collision. Then we may write:

$$\left.\begin{aligned} \text{distance moved by planes} &= x_{n+1} - x_n = -V \Delta t_n \\ \text{distance moved by ball} &= x_{n+1} + x_n = v_{n+1} \Delta t_n. \end{aligned}\right\} \tag{9.8}$$

The action between successive collisions is I_n, where

$$\pi I_n = v_n x_n \tag{9.9}$$

so that, on using (9.8), we obtain

$$I_{n+1} = \left(1 + \frac{2z_n^2}{1 + 3z_n} \right) I_n \quad \left(z_n = \frac{V}{v_0 + 2nV} \right). \tag{9.10}$$

This relation between I_{n+1} and I_n may be used to calculate I_n given I_0 and z_0. In table 9.1 we show I_n for various n and $V = 0.01$, $v_0 = 1$, $I_0 = 1/\pi$; also shown are the energy and amplitude of the motion, x_n.

Table 9.1

n	$I_n \pi$	$E_n = \frac{1}{2} v_n^2$	x_n
0	1	0.50	1
10	1.0017	0.72	0.84
50	1.0050	2.00	0.50
100	1.0066	4.50	0.34
500	1.0090	60.5	0.09
1000	1.0094	220.5	0.05

In this example we see that the action is almost constant whilst the energy and amplitude of the motion change considerably. For example, after 100 collisions the energy, amplitude and action change by factors of *9, 3* and *1.007* respectively. We can use the near constancy of the action to estimate the energy and other dynamical quantities, such as frequency, when direct estimates are more difficult.

If the planes move apart with constant velocity, at each collision the ball loses energy; so, after a finite number of collisions, it will be moving too slowly to catch the moving plane and the motion will change from being bound to being unbound. Nevertheless, whilst the motion is bound, a similar analysis shows that the action is almost constant whereas both energy and amplitude vary considerably, as in exercise 9.9.

An elementary model of a gas comprises many non-interacting atoms bouncing elastically between the walls of a container. The adiabatic invariance of the action gives rise to the gas laws. For example in a cubic box of side L and volume $V = L^3$, we may suppose every atom to be moving with speed v perpendicular to the plane faces of the box, with equal numbers moving in each direction. Then the invariance of the action under slow changes in L gives $vL = $ constant, and, since the temperature, T, is proportional to the kinetic energy, $TV^{\frac{2}{3}} = $ constant. This gas law is valid if the volume of the gas changes slowly, that is, if the walls move slowly by comparison to the speed of the atoms; this condition is often violated in explosions and shock waves, and then the gas laws also break down, as we should expect from the theory of adiabatic invariance.

9.3   The linear oscillator with a slowly changing frequency

Consider the Hamiltonian for a particle moving under the influence of a linear attractive force of slowly changing magnitude. It is convenient to introduce a scaled time $\lambda = \epsilon t$, where ϵ is a small constant chosen so that the force constant $\omega(\lambda)$ only changes significantly in a scaled time of order $\lambda = 1$, corresponding to an actual time of order ϵ^{-1}. Then

$$\frac{d\omega}{dt} = \epsilon \frac{d\omega}{d\lambda}, \tag{9.11}$$

where $|\, d\omega/d\lambda \,|$ is of order unity or less. The Hamiltonian then has the form

$$H = p^2/2m + \tfrac{1}{2} m\, \omega(\lambda)^2\, q^2. \tag{9.12}$$

First consider the trivial case where λ is constant. Then ω is constant and, as shown in example 4.2, the motion is simple harmonic with period $2\pi/\omega$; for convenience we call this the unperturbed motion. For this case we show in example 7.5 that the system has angle–action variables (θ, I), where

$$q = \left[\frac{2I}{m\,\omega(\lambda)} \right]^{\frac{1}{2}} \sin\theta,$$

$$p = [2mI\omega(\lambda)]^{\frac{1}{2}} \cos\theta, \tag{9.13}$$

with generating function

$$S_1(\theta, q; \lambda) = \tfrac{1}{2} m\, \omega(\lambda)\, q^2 \cot\theta. \tag{9.14}$$

When λ changes slowly then so does $\omega(\lambda)$; the change in ω is called *adiabatic* if ω changes relatively little during one period of the unperturbed motion, that is

$$\text{period} \times \left| \frac{d\omega}{dt} \right| \ll \omega, \tag{9.15}$$

or

$$| \epsilon \, \omega'(\lambda) | \ll \omega^2, \tag{9.16}$$

where $\omega'(\lambda) = d\omega/d\lambda$. Notice that this inequality breaks down, as does the following theory, if ω becomes small. In general, angle–action variables do not exist for time-dependent systems, but in this case of slowly varying parameters the conjugate variables (θ, I) are the best representation with which to understand and approximate the motion.

The transformation $(q, p) \to (\theta, I)$ generated by $S_1(\theta, q; \lambda)$ is now time-dependent so that the theory of section 6.7 can be used to find the Hamiltonian in the (θ, I) representation:

$$K(\theta, I, \lambda) = I\omega(\lambda) + \partial S_1 / \partial t$$

$$= I\omega(\lambda) + \frac{\epsilon I \omega'(\lambda)}{2\omega(\lambda)} \sin 2\theta. \tag{9.17}$$

The equations of motion are

$$\dot{\theta} = \omega(\lambda) + \frac{\epsilon \omega'(\lambda)}{2\omega(\lambda)} \sin 2\theta, \tag{9.18}$$

$$\dot{I} = - \frac{\epsilon \omega'(\lambda)}{\omega(\lambda)} \cos 2\theta \quad (\lambda = \epsilon t). \tag{9.19}$$

These equations cannot be solved exactly. Soon we shall find an approximate solution, but first we examine the equations themselves in order to obtain a qualitative picture of the motion. The essential point is that θ is rapidly varying while I and ω are slowly varying.

First consider $\theta(t)$: because $| \epsilon \omega'/\omega^2 | \ll 1$ we have

$$\dot{\theta} \simeq \omega(\lambda), \tag{9.20}$$

so that θ increases approximately linearly with time, increasing by π in a time interval of approximately π/ω. Upon this motion the other small term of equation (9.18) superimposes small oscillations with a period of approximately π/ω and an amplitude of order ϵ.

The behaviour of $I(t)$ is quite different. The essential feature of equation (9.19) is that the right-hand side is periodic in θ with zero mean value. Thus when θ changes by π, $I(t)$ undergoes change of order ϵ^2, although during this time interval it oscillates with an amplitude of order ϵ. We describe this behaviour by saying that the mean rate of change of I is of order ϵ^2.

It is this very slow mean rate of change of $I(t)$ that makes the (θ, I) representation the best for understanding the motion. Furthermore, the equations of motion in this representation are in a form suited to a perturbation expansion as described in chapter 8. On the other hand the equations of motion in the (q, p) representation,

$$\frac{dq}{dt} = p/m, \quad \frac{dp}{dt} = -m\omega(\lambda)^2 q \quad (\lambda = \epsilon t), \tag{9.21}$$

are not in a form suited to a perturbation expansion.

To apply perturbation theory we write $(\theta(t), I(t))$ as a power series in ϵ:

$$\theta(t) = \theta^{(0)}(t) + \epsilon\, \theta^{(1)}(t) + O(\epsilon^2) \tag{9.22}$$

$$I(t) = I^{(0)}(t) + \epsilon\, I^{(1)}(t) + O(\epsilon^2). \tag{9.23}$$

Substituting these into the equations of motion, (9.18) and (9.19), and equating powers of ϵ gives

$$\dot{\theta}^{(0)} = \omega(\lambda), \quad \dot{I}^{(0)} = 0, \tag{9.24}$$

having solutions

$$\theta^{(0)} = \int_0^t dt\, \omega(\lambda) + \theta_0, \quad I^{(0)} = I_0, \tag{9.25}$$

where (θ_0, I_0) are the initial values of the angle–action variables.

Similarly the equations for the first-order corrections are found to be

$$\dot{\theta}^{(1)} = \frac{\omega'}{2\omega}\, \sin 2\theta^{(0)}, \quad \theta^{(1)}(0) = 0; \tag{9.26}$$

$$\dot{I}^{(1)} = \frac{-\omega'}{\omega}\, I_0 \cos 2\theta^{(0)}, \quad I^{(1)}(0) = 0. \tag{9.27}$$

These equations may be integrated directly. For example, the right-hand side of (9.27) can be expressed as the sum of two parts:

$$\frac{\omega'}{\omega} \cos 2\theta^{(0)} = \frac{d}{dt}\left(\frac{\omega'}{2\omega^2} \sin 2\theta^{(0)}\right) - \frac{\epsilon}{2}\left(\frac{\omega'}{\omega^2}\right)' \sin 2\theta^{(0)}. \tag{9.28}$$

The second term on the right, being of order ϵ, may be ignored. The first term can be integrated directly so that equation (9.23) becomes

$$I(t) = I_0\left[1 - \frac{\epsilon}{2}\left(\frac{\omega'}{\omega^2} \sin 2\theta^{(0)} - \frac{\omega_0'}{\omega_0^2} \sin 2\theta_0\right)\right] + O(\epsilon^2), \tag{9.29}$$

where $\omega_0 = \omega(0)$. Similarly

$$\theta(t) = \theta^{(0)}(t) - \frac{\epsilon}{4}\left(\frac{\omega'}{\omega^2} \cos 2\theta^{(0)} - \frac{\omega_0'}{\omega_0^2} \cos 2\theta_0\right) + O(\epsilon^2). \tag{9.30}$$

In both cases the first-order correction is small by virtue of inequality (9.16).

As $\theta^{(0)}$ increases by π,

$$\frac{\omega'}{2\omega^2} \sin 2\theta^{(0)}$$

changes by a quantity of order ϵ, so that $I(t)$ changes by a quantity of order ϵ^2. The behaviour of (θ, I) is sketched in figure 9.1.

Fig. 9.1 Sketch showing the variation of $\theta(t)$ and $I(t)$ over the time interval $(0, \epsilon^{-1})$. The left-hand side shows $\theta(t)$ which consists of a monotonic variation $\theta^{(0)}(t)$, equation (9.25), on which are super-imposed the rapid oscillations $\theta^{(1)}(t)$, equation (9.26). The right-hand side shows $I(t)$, which consists of the constant $I^{(0)}$, equation (9.25), on which are superimposed the rapid oscillations of $I^{(1)}$, equation (9.27).

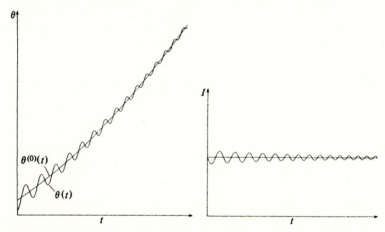

The behaviour in the (q, p) representation is more complicated than this. An approximate solution in this representation, having errors of order ϵ, is obtained by substituting $(\theta^{(0)}, I^{(0)})$ for (θ, I) in equation (9.13):

$$q(t) = \left[\frac{2I_0}{m\omega(\lambda)}\right]^{\frac{1}{2}} \sin\left(\int_0^t dt\, \omega + \theta_0\right),$$

$$p(t) = [2mI_0\, \omega(\lambda)]^{\frac{1}{2}} \cos\left(\int_0^t dt\, \omega + \theta_0\right).$$

(9.31)

During the time interval $(t, t + 2\pi/\omega)$ the phase curve in this representation given by (9.31) is an approximate ellipse. The exact phase curve differs from this by a quantity of order ϵ.

In figure 9.2 we show segments of the phase curve described by (9.31) for the particular case

$$\omega = (1 + \epsilon t)\Omega \quad (\Omega = \text{constant})$$

(9.32)

for the time intervals $\Delta t_1 = (0, 2\pi/\Omega)$, $\Delta t_2 = (1/2\epsilon, 1/2\epsilon + 4\pi/3\Omega)$ and $\Delta t_3 = (1/\epsilon, 1/\epsilon + \pi/\Omega)$.

Each of these segments is an approximate ellipse, each of which is significantly different from others. In the unperturbed case, $\epsilon = 0$, the phase curve is an ellipse approximately the shape of curve 1 in figure 9.2. So, for $\epsilon \neq 0$, the phase curve in the (q, p) representation differs significantly from the unperturbed phase curve in a time $(0, \epsilon^{-1})$. Note that the area of the curves remains approximately the same. This is because $I(t)$ is nearly constant and $2\pi I(t)$ is the area.

Fig. 9.2 Segments of the phase curve in the (q, p) representation during different time intervals when $\omega(t)$ is given by equation (9.32). Curve (1) is during the interval $(0, 2\pi/\Omega)$, curve (2) $(1/2\epsilon, 1/2\epsilon + 4\pi/3\Omega)$ and curve (3) $(1/\epsilon, 1/\epsilon + \pi/\Omega)$. Note that the approximately elliptical phase curve of each segment is not quite closed.

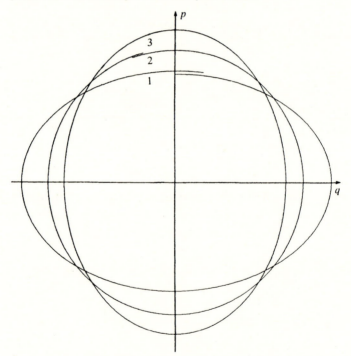

The approximate solution for $I(t)$, equation (9.29), shows that

$$|I(t) - I(0)| < K\epsilon \tag{9.33}$$

for some constant K, but it does not establish for how long this inequality is valid. This is found by writing the equation of motion for $I(t)$, (9.19), in the form

$$\dot{I} = -\epsilon \, \frac{\mathrm{d}}{\mathrm{d}t} \left[\frac{I\omega'}{2\omega^2} \, \sin 2\theta \right] + \epsilon^2 M(t), \tag{9.34}$$

where

$$M(t) = M(\theta, I, \lambda) = \tfrac{1}{2} I \left(\frac{\omega'}{\omega} \right)' \sin 2\theta. \tag{9.35}$$

On integrating (9.34) we obtain

$$I(t) - I_0 = -\epsilon \left(\frac{I\omega'}{2\omega^2} \, \sin 2\theta - \frac{I_0 \omega_0'}{2\omega_0^2} \, \sin 2\theta_0 \right) + \epsilon^2 \int_0^t \mathrm{d}t \, M(t), \tag{9.36}$$

but $M(t)$ is a bounded function of time

$$M(t) \leqslant M_1 = \max \left[\frac{I}{2} \left(\frac{\omega'}{\omega} \right)' \right] \tag{9.37}$$

and similarly

$$\left| \frac{I(t) \, \omega'}{2\omega^2} \, \sin 2\theta \right| \leqslant M_2 = \max \left(\frac{I\omega'}{2\omega^2} \right), \tag{9.38}$$

for some constants M_1 and M_2. Thus the difference between $I(t)$ and its initial value is bounded by

$$|I(t) - I_0| \leqslant 2 \, \epsilon M_2 + \epsilon^2 t M_1, \tag{9.39}$$

so that

$$|I(t) - I_0| < K\epsilon \qquad (0 < t < \epsilon^{-1}), \tag{9.40}$$

meaning that $I(t)$ remains close to its initial value for all times in the interval $(0, \epsilon^{-1})$.

9.4 General adiabatic theory

We have shown explicitly that for some systems the action is approximately constant when the Hamiltonian changes adiabatically. We now show that this is true in general if certain conditions are satisfied.

Consider the Hamiltonian

$$H = H(q, p, \lambda), \tag{9.41}$$

where λ is a parameter. If λ is constant we suppose that the motion is bounded so that there exist different angle–action variables for each λ. The adiabatic variation of H may then be expressed by putting

$$\lambda(t) = \epsilon t \quad (0 < \epsilon \ll 1), \tag{9.42}$$

where ϵ is chosen so that the Hamiltonian only changes significantly in a scaled time of order $\lambda = 1$, corresponding to an actual time of order ϵ^{-1}. Then

$$\frac{\partial H}{\partial t} = \epsilon \frac{\partial H}{\partial \lambda}, \qquad (9.43)$$

where $|\partial H/\partial \lambda|$ is of order unity or less.

We call a function $F(q, p, \lambda)$ an *adiabatic invariant* of the system if, for any ϵ satisfying $0 < \epsilon \ll 1$,

$$F(t) = F(q(t), p(t), \lambda(t)) \qquad (9.44)$$

varies little during the time $0 < t < \epsilon^{-1}$. To be precise, if for any $\eta > 0$ it is possible to find an ϵ_0 such that for any $0 < \epsilon < \epsilon_0$, $\lambda(t)$ given by (9.42) and

$$|F(t) - F(0)| < \eta \qquad (0 < t < \epsilon^{-1}), \qquad (9.45)$$

then F is called an *adiabatic invariant*.

Notice that an adiabatic invariant is approximately constant for a given, limited, time interval. Sometimes, but not always, it is approximately constant for a longer time interval, even for ever.

Adiabatic invariance of the action variable
This general analysis could be omitted at a first reading, in which case the reader should turn to example 9.1.

We suppose that, for constant λ, the Hamiltonian (9.41) has the angle–action variables (θ, I) with the canonical transformation $(q, p) \to (\theta, I)$ defined by the generating function $S_1(\theta, q; \lambda)$ depending parametrically upon λ.

If λ varies with time the generating functions also depends upon time: the transformation $(q, p) \to (\theta, I)$ generated by S_1 is still canonical but the variables (θ, I) are no longer the angle–action variables for the system. The new Hamiltonian is obtained from the theory of time-dependent transformations of section 6.7, equation (6.70), using equation (9.42):

$$K(\theta, I, \lambda) = H(I, \lambda) + \epsilon \frac{\partial S_1}{\partial \lambda}, \qquad (9.46)$$

where

$$H(I, \lambda) = H(q(\theta, I, \lambda), p(\theta, I, \lambda), \lambda), \qquad (9.47)$$

is by definition a function of I and λ only.

Hamilton's equations of motion are then

$$\dot{\theta} = \omega(I, \lambda) + \epsilon R_1(\theta, I, \lambda), \qquad (9.48)$$

$$\dot{I} = -\epsilon R_2(\theta, I, \lambda), \qquad (9.49)$$

where

$$\omega(I, \lambda) = \frac{\partial H}{\partial I}(I, \lambda) \qquad (9.50)$$

and

$$R_1 = \frac{\partial^2 S_1}{\partial \lambda \partial I} \ , \qquad R_2 = \frac{\partial^2 S_1}{\partial \lambda \partial \theta} \ . \tag{9.51}$$

These equations of motion are similar in form to those obtained for the adiabatically changing linear oscillator, equation (9.18) and (9.19), where

$$R_1 = \frac{\omega' \sin 2\theta}{2\omega} \ , \qquad R_2 = \frac{I\omega'}{\omega} \cos 2\theta. \tag{9.52}$$

In that case the simplicity of the motion in the (θ, I) representation was due to both R_1 and R_2 being periodic functions of θ, because R_2 had zero mean value when averaged over θ and because

$$\omega(I, \lambda) \gg \epsilon \max |R_1|. \tag{9.53}$$

If R_1 and R_2 always have these properties then the analysis used for the linear oscillator, in particular equation (9.34) may be slightly modified to show that the action is an adiabatic invariant.

First we show that R_1 and R_2 have the necessary periodicity properties. It was shown in section 7.4 that $S_1(\theta, q; \lambda)$ is periodic in θ and hence so are R_1 and R_2. The mean value of R_2 is

$$\frac{1}{2\pi} \int_0^{2\pi} d\theta \, R_2 = \frac{1}{2\pi} \int_0^{2\pi} d\theta \, \frac{\partial^2 S_1}{\partial \lambda \partial \theta} \ , \tag{9.54}$$

where, in the integral, I and λ are held fixed. The right-hand side may be written

$$\frac{1}{2\pi} \frac{\partial}{\partial \lambda} \int_0^{2\pi} d\theta \, \frac{\partial S_1}{\partial \theta} = \frac{1}{2\pi} \frac{\partial}{\partial \lambda} \, [S_1]_{\theta=0}^{\theta=2\pi} = 0 \tag{9.55}$$

since S_1 is periodic in θ. Hence the mean value of R_2 is zero.

To establish the adiabatic invariance of I it is convenient to write the right-hand side of (9.49) as a sum of a derivative of some function with respect to time and a quantity of order ϵ^2, as in the linear oscillator example, equation (9.34). To this end it is useful to define

$$R_3(\theta, I, \lambda) = \int_0^\theta d\theta' \, R_2(\theta', I, \lambda) \tag{9.56}$$

so that R_3 is also periodic, because the mean of R_2 is zero. Then

$$\frac{d}{dt} \left(\frac{R_3}{\omega} \right) = \frac{1}{\omega} \frac{\partial R_3}{\partial \theta} \dot{\theta} + \frac{\partial}{\partial I} \left(\frac{R_3}{\omega} \right) \dot{I} + \frac{\partial}{\partial \lambda} \left(\frac{R_3}{\omega} \right) \dot{\lambda} \tag{9.57}$$

which can be rearranged to give

$$R_2 = \frac{\mathrm{d}}{\mathrm{d}t}\left(\frac{R_3}{\omega}\right) - \epsilon M(t),$$ (9.58)

where

$$M(t) = \frac{R_1 R_2}{\omega} - R_2 \frac{\partial}{\partial I}\left(\frac{R_3}{\omega}\right) + \frac{\partial}{\partial \lambda}\left(\frac{R_3}{\omega}\right).$$ (9.59)

Substituting this relation for R_2 into the equation of motion for I, equation (9.49), gives

$$\dot{I} = -\epsilon \frac{\mathrm{d}}{\mathrm{d}t}\left(\frac{R_3}{\omega}\right) + \epsilon^2 M(t).$$ (9.60)

Since R_1, R_2 and R_3 are periodic and $\omega \neq 0$, $M(t)$ is bounded,

$$|M(t)| \leqslant M_1,$$ (9.61)

as is R_3/ω,

$$\left|\frac{R_3}{\omega}\right| \leqslant M_2,$$ (9.62)

for some constants M_1 and M_2. Thus the difference between $I(t)$ and its initial value is bounded by

$$|I(t) - I_0| \leqslant 2\epsilon M_2 + \epsilon^2 t M_1,$$ (9.63)

so that for some constant K

$$|I(t) - I_0| \leqslant K\epsilon \quad (0 < t < \epsilon^{-1}),$$ (9.64)

meaning that $I(t)$ remains close to its initial value for all times in the interval $(0, \epsilon^{-1})$.

This very simple result is also very powerful. It often enables useful information to be obtained about the motion of a time-dependent system from the much simpler motion of a related time-independent system. This is illustrated in the next example.

Example 9.1

A particle of mass m moves in the potential

$$V(q) = U \tan^2 \alpha q.$$

How do the energy, amplitude and frequency of the motion change if α increases slowly?

In problems concerning adiabatic invariance we always consider the related time-independent problem first. This is done in example 7.2 and we see from equation (7.18) that the energy and action are related by

$$E = \alpha I [\alpha I + 2(2mU)^{\frac{1}{2}}]/2m$$

and from equation (7.19) that the frequency and action are related by

$$\omega = \alpha[\alpha I + (2mU)^{\frac{1}{2}}]/2m.$$

During an adiabatic change of the Hamiltonian the action remains constant; the value of the Hamiltonian is not constant, so the energy varies, however the above relation between the energy, α and I remains approximately true so that the energy at time t is simply

$$E(t) \simeq \alpha(t)I[\alpha(t)I + 2(2mU)^{\frac{1}{2}}]/2m.$$

Similarly, we see from equation (9.48) that the frequency of the motion is simply

$$\omega(t) = \alpha(t)[\alpha(t)I + (2mU)^{\frac{1}{2}}]/m.$$

In this case both the energy and frequency of the motion increase with $\alpha(t)$.

The turning point for constant α is given in example 7.2. For a slowly varying α we have

$$q_1(t) = \frac{1}{\alpha(t)} \tan^{-1} \left[\left(\frac{E(t)}{U} \right)^{\frac{1}{2}} \right].$$

For small α, $q_1(t)$ is almost constant, but for large α the amplitude is inversely proportional to $\alpha(t)$. When $\alpha(t)$ is large and slowly varying the motion is close to that of the elastic ball bouncing between two slowly moving planes discussed in section 9.2.

9.5   Motion in a rapidly oscillating field: fast perturbations

Let us now consider the effect of adding a rapidly varying field to a Hamiltonian $H_0(q, p)$. To be specific, let the variation be sinusoidal, so that

$$H(q, p) = H_0(q, p) + V(q) \sin \omega t, \tag{9.65}$$

and the period $2\pi/\omega$ is small by comparison with the time during which the unperturbed system changes significantly.

In order to understand the qualitative features of the motion we consider a free particle perturbed by a rapidly oscillating uniform field: the Hamiltonian is

$$H(q, p) = p^2/2m + qF \sin \omega t, \tag{9.66}$$

F being a constant, and the equation of motion is

$$m\ddot{q} = F \sin \omega t. \tag{9.67}$$

This may be integrated directly to give

$$q(t) = q_0(t) + \xi(t), \tag{9.68a}$$

$$\xi(t) = - \frac{F \sin \omega t}{m\omega^2}, \tag{9.68b}$$

where $q_0(t)$ is the unperturbed motion. In this case the motion consists of the unperturbed motion on which are superimposed small rapid oscillations, $\xi(t)$, of period $2\pi/\omega$ and zero mean value.

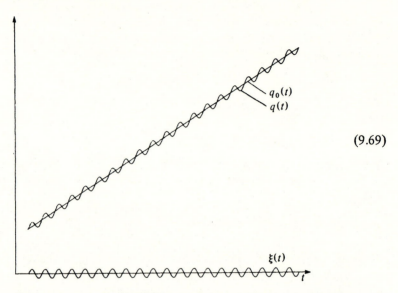

$$(9.69)$$

The mean motion, that is the motion averaged over one period of the forcing term, is $q_0(t)$. Notice that, since ω is large, the perturbed and unperturbed motions are close even when the magnitude of the oscillating force, F, is large. This is because the force changes direction so rapidly that the system has an insufficient time to respond significantly. By contrast to the previous section the rapid oscillations here are externally produced.

The momentum is given by

$$p(t) = m\dot{q} = p_0 + \eta, \qquad (9.70a)$$

$$\eta(t) = -\frac{F \cos \omega t}{\omega}, \qquad (9.70b)$$

so that the perturbations of it are a factor ω larger than those of the position.

This simple example suggests that an approximate solution to the equations of motion for the general Hamiltonian (9.65) may be formed by writing

$$q(t) = \bar{q}(t) + \xi(t),$$
$$p(t) = \bar{p}(t) + \eta(t), \qquad (9.71)$$

where $(\bar{q}, \bar{p})$ are slowly changing, (ξ, η) are small, periodic with period $2\pi/\omega$ and with zero mean value. By analogy with the simple case above, we would expect the following quantities

$$\ddot{\xi}, \quad \omega\dot{\xi}, \quad \omega^2\xi, \quad \omega\eta/m \tag{9.72}$$

to have similar orders of magnitude.

Substituting (9.71) into the Hamiltonian (9.65) and expanding to the appropriate order gives the following equations of motion:

$$\dot{\bar{q}} + \dot{\xi} = \frac{\partial H_0}{\partial \bar{p}} + \eta \; \frac{\partial^2 H_0}{\partial \bar{p}^2} + \tfrac{1}{2} \; \eta^2 \; \frac{\partial^3 H_0}{\partial \bar{p}^3}, \tag{9.73a}$$

$$\dot{\bar{p}} + \dot{\eta} = - \frac{\partial H_0}{\partial \bar{q}} - \eta \; \frac{\partial^2 H_0}{\partial \bar{q} \partial \bar{p}} - \frac{\partial V}{\partial \bar{q}} \sin \omega t - \frac{\partial^2 V}{\partial \bar{q}^2} \xi \sin \omega t$$

$$- \tfrac{1}{2} \; \eta^2 \; \frac{\partial^3 H_0}{\partial \bar{q} \partial \bar{p}^2}. \tag{9.73b}$$

In these equations the first terms on each side are slowly varying and the rest rapidly varying. On taking the mean over a fast oscillation of period $2\pi/\omega$, and assuming the variation of $(\bar{q}, \bar{p})$ negligible in this period, we obtain the averaged equations for the mean motion $(\bar{q}, \bar{p})$

$$\dot{\bar{q}} = \frac{\partial H_0}{\partial \bar{p}} + \frac{1}{2} \frac{\partial^3 H_0}{\partial \bar{p}^3} \langle \eta^2 \rangle, \tag{9.74a}$$

$$\dot{\bar{p}} = - \frac{\partial H_0}{\partial \bar{q}} - \frac{\partial^2 V}{\partial \bar{q}^2} \langle \xi \sin \omega t \rangle - \tfrac{1}{2} \langle \eta^2 \rangle \frac{\partial^3 H_0}{\partial \bar{q} \partial \bar{p}^2}. \tag{9.74b}$$

The small quantities (ξ, η) are found by substituting (9.74) back into (9.73) and retaining only the largest terms,

$$\dot{\xi} = \eta \; \frac{\partial^2 H_0}{\partial \bar{p}^2} \tag{9.75a}$$

$$\dot{\eta} = - \frac{\partial V}{\partial \bar{q}} \sin \omega t. \tag{9.75b}$$

These may be integrated directly if it is assumed that $(\bar{q}, \bar{p})$ vary negligibly during one fast oscillation:

$$\xi = \frac{1}{\omega^2} \; \frac{\partial V}{\partial \bar{q}} \; \frac{\partial^2 H_0}{\partial \bar{p}^2} \sin \omega t, \tag{9.76a}$$

$$\eta = \frac{1}{\omega} \; \frac{\partial V}{\partial \bar{q}} \cos \omega t. \tag{9.76b}$$

Thus equations (9.74) may be written in the Hamiltonian form

$$\dot{\bar{q}} = \frac{\partial K}{\partial \bar{p}} (\bar{q}, \bar{p}), \quad \dot{\bar{p}} = - \frac{\partial K}{\partial \bar{q}} (\bar{q}, \bar{p}) \tag{9.77}$$

where the time-independent Hamiltonian is

$$K(\bar{q},\bar{p}) = H_0(\bar{q},\bar{p}) + \frac{1}{4\omega^2} \left(\frac{\partial V}{\partial \bar{q}} \right)^2 \frac{\partial^2 H_0}{\partial \bar{p}^2} . \tag{9.78}$$

This is the *mean motion Hamiltonian* and it differs from the unperturbed Hamiltonian by a quantity of order ω^{-2}. The actual motion consists of small, fast, periodic oscillations about this mean.

In the particular case where the unperturbed Hamiltonian has the form

$$H_0 = p^2/2m + V_0(q), \tag{9.79}$$

the modified Hamiltonian may be expressed in terms of an effective potential:

$$K(\bar{q},\bar{p}) = \bar{p}^2/2m + V_{\text{eff}}(\bar{q}), \tag{9.80a}$$

where

$$V_{\text{eff}}(\bar{q}) = V_0(\bar{q}) + \frac{1}{4m\omega^2} \left(\frac{\partial V}{\partial \bar{q}} \right)^2 . \tag{9.80b}$$

The mean motion is the same as if the unperturbed potential V_0 were augmented by a potential of magnitude proportional to the square of the oscillating force. In particular, for a uniform oscillating field $V(q) = qF$, the mean motion is the same as the unperturbed motion, since K and H differ only by a constant. In that case the small oscillations about the mean motion are

$$\xi = \frac{F}{m\omega^2} \sin \omega t , \quad \eta = \frac{F}{\omega} \cos \omega t. \tag{9.81}$$

Example 9.2
Consider the motion of a vertical pendulum comprising a light stiff rod of length l with a mass m fixed at its end and whose point of support oscillates vertically with frequency ω and amplitude a.

The Hamiltonian for this system is obtained in example 5.7, equation (5.38) with $\gamma(t) = a \sin \omega t$:

$$H = \frac{p^2}{2ml^2} - mgl(1 + \frac{a\omega^2}{g} \sin \omega t) \cos \psi \tag{9.82}$$

The Hamiltonian for the mean motion is obtained using equation (9.78):

$$K = \frac{\bar{p}^2}{2ml^2} - mgl (\cos \psi - k \sin^2 \psi), \tag{9.83}$$

where

$$k = \frac{a^2 \omega^2}{4gl} .$$ (9.84)

The fixed points of the mean motion are at the roots of

$$\frac{\mathrm{d} V_{\mathrm{eff}}}{\mathrm{d} \psi} = 0,$$ (9.85)

where

$$V_{\mathrm{eff}} = - mgl (\cos \psi - k \sin^2 \psi).$$ (9.86)

These are at $\psi = 0, \pi$ and $\cos \psi = -1/2k$.

At $\psi = 0$, $\mathrm{d}^2 V_{\mathrm{eff}}/\mathrm{d}\psi^2 > 0$ so that this is always a stable fixed point, as expected from physical considerations.

At $\psi = \pi$

$$\frac{\mathrm{d}^2 V_{\mathrm{eff}}}{\mathrm{d}\psi^2} = mgl (2k - 1)$$ (9.87)

so that this is a stable fixed point if

$$a^2 \omega^2 > 2gl$$ (9.88)

otherwise it is unstable. Thus, for fast enough oscillations, the pendulum is stable when its centre of gravity is vertically above the point of support.

At $\cos \psi = -1/2k$

$$\frac{\mathrm{d}^2 V_{\mathrm{eff}}}{\mathrm{d}\psi^2} = mgl \left(\frac{1}{2k} - 2k \right)$$ (9.89)

so that for real ψ this fixed point is always unstable.

Exercises for chapter 9
Appendix 2 contains some of the integrals required for these exercises.

Exercises on adiabatic theory

(1) A vertical pendulum consists of a mass M swinging on a light string of length $l(t)$ which varies slowly with time. Write down the Hamiltonian for the system as it executes small oscillations and use the principle of adiabatic invariance to show that

$$\frac{\omega_2}{\omega_1} \simeq \left(\frac{l_1}{l_2} \right)^{\frac{1}{2}},$$

where ω_i is the frequency of the system when the length of the string is l_i.

(2) An ice skater S is freely swung around a fixed point, 0, on horizontal smooth ice. Write down the Hamiltonian if the distance $R(t)$ from 0 to S is a given function of t.

  (a) Use the principle of adiabatic invariance to show that if $R(t)$ changes slowly from R_0 to αR_0, then the skater's kinetic energy changes from E_0 to approximately E_0/α^2.

  (b) Show that in this example the action is an exact invariant. In the particular case $R(t) = (1 + \epsilon t)R_0$, where ϵ is a small positive constant, find the motion explicitly and verify that the above approximation to the change in the kinetic energy is accurate to order ϵ for all time.

(3) Derive equation (9.30) of the text.

(4) Show that, if (θ, I) satisfy the equations

$$\dot\theta = \omega(I) \neq 0,$$

$$\dot I = \epsilon g(\theta) \quad (0 < \epsilon \ll 1),$$

where $g(\theta)$ is a periodic function of θ with period 2π, then

$$|I(t) - M(t)| < K\epsilon \quad (0 < t < \epsilon^{-1}),$$

K being a constant, and

$$M(t) = I(0) + \epsilon t \left[\frac{1}{2\pi} \int_0^{2\pi} d\theta \, g(\theta) \right].$$

Demonstrate this explicitly in the following cases:

  (a) $\omega = $ constant, $\quad g(\theta) = \sin\theta$,

  (b) $\omega = $ constant, $\quad g(\theta)$ is a periodic function,

  (c) $\omega = I\alpha$, α constant, $g(\theta) = \cos\theta$.

Is the restriction $0 < t < \epsilon^{-1}$ necessary in all of these cases?

(5) A particle of mass m moves in the potential

$$V(q) = A \tan^2(\alpha q) \quad (|q| \leqslant \pi/2\alpha),$$

A and α being positive constants. Sketch the potential and phase curves and give a qualitative description of the motion. Also find the angle-action variables and hence solve for $q(t)$. If A changes slowly to kA $(k > 0)$, find the consequent changes to the energy and frequency.

(6) How does the energy of a particle of mass m moving in a potential $V(q)$ change when the parameters of the potential change slowly for the following cases:

  (a) $V(q) = -A/\cosh^2\alpha q$.

  (b) $V(q) = Aq^{2n}$,

where A and α are positive constants and n is an integer?

(7) A particle of mass M moves in the potential

$$V(q) = A(e^{-2\alpha q} - 2e^{-\alpha q}),$$

where α is a positive constant. If A is slowly changed to $2A$ show that the difference between the initial and final frequencies of the motion is

$$\omega_f - \omega_i = \alpha \left(\frac{A}{M}\right)^{\frac{1}{2}} (2 - \sqrt{2}).$$

(8) A particle moves smoothly up and down an inclined plane AB, as shown in the diagram, and is elastically reflected by a wall at A. How does the maximum height the particle reaches change when the angle α changes slowly?

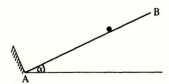

(9) If the planes of section 9.2 are moving apart with constant velocity V, show that

$$I_{n+1} = I_n \left(1 + \frac{2z_n^2}{1 - 3z_n}\right).$$

Find z_n and show that there are about $v_0/2V$ collisions with the moving plane before the motion changes character.

(10) Two elastic particles of masses μM and M, with $\mu \ll 1$, move along the straight line OA.

At O the particle μM is reflected elastically by a stationary wall. If the collisions between the two masses are also elastic, show that the action of μM after the nth collision with M is given by

$$I_{n+1} = \left[1 + \frac{2(z_n^2 + \mu)}{1 - 3\mu + z_n(3 - \mu)}\right] I_n,$$

$$z_{n+1} = \frac{2\mu + (1 - \mu)z_n}{1 - \mu - 2z_n} \qquad (z_n = V_n/v_n),$$

where the notation is the same as in section (9.2), V being the speed of M and v that of μM.

(11) Assuming that the action of the mass μM of the previous question is an adiabatic invariant, show that the mean motion of M satisfies the equation

$$\frac{dV}{dt} = \frac{\mu I^3}{X^3} ,$$

X being the distance of M from 0. Hence show that

$$X \simeq \left(\frac{\mu I^3}{c} + c(t - t_0)^2 \right)^{\frac{1}{2}}$$

c and t_0 being constants.

(12) A small ball is bouncing up and down on an elastic plate in a lift. What is the change in the maximum height the ball reaches if the acceleration of the lift is slowly changed? (Hint: You may assume the result given at the end of chapter 5 that the vertical acceleration has the same effect as a time-varying gravitational field.)

(13) A particle of mass m moves freely in the space between two fixed surfaces as shown in the diagram.

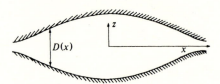

If the particle bounces off both planes elastically, if $|\dot{z}| \gg |\dot{x}|$ and if $D(x)$ is the separation of the surfaces at a point x, use the conservation of energy and the principle of adiabatic invariance to show that the motion in the x-direction is described approximately by

$$\dot{x}^2 = \frac{2E}{m} - \left(\frac{C}{D(x)} \right)^2 ,$$

C being a constant and E the total energy. Deduce that, if $D(x)$ is sufficiently small, the particle is trapped in a finite range of x.

(14) A particle of mass m bounces elastically between the two fixed surfaces $z = 0$ and $z = a \cosh \alpha x / \cosh 2\alpha x$. The particle starts at the origin in the zx-plane moving at a small angle β to the z-axis. Using the results of the previous exercise determine the maximum value of x and the approximate period of the motion along the x-axis.

(15) Find the connection between the volume and the pressure of a 'gas'

consisting of particles moving parallel to the edges and inside an elastic cube, when the size of the cube changes slowly.

Exercises on rapidly oscillating Hamiltonians.

(16) A particle of mass m is attracted towards a point P with coordinate ξ on the real axis with a force $-m\Omega^2 X$, where X is the distance between P and the particle. If P oscillates about the origin with a motion described by

$\xi = a \sin \omega t$,

where $\omega \gg \Omega$, show that the mean motion of the particle is a simple harmonic oscillation about the origin with frequency Ω.

(17) The Hamiltonian for a particle of mass m and electrical charge e moving along the positive x-axis acted on by two charges $\mp e$ at $x = \pm a$ respectively is

$$H = p^2/2m - \mu e^2 \left(\frac{1}{x-a} - \frac{1}{x+a} \right),$$

where μ is a constant. Show that at large distances, $x \gg a$, the particle moves in the approximate attractive mechanical potential

$V(x) = -2\mu e^2 a/x^2$.

If the two charges execute rapid oscillations of the form $a(t) = \pm a_0 \sin \omega t$ show that the mean mechanical potential at large distances is

$$\overline{V}(x) = \frac{1}{m} \left(\frac{2\mu e^2 a_0}{\omega} \right)^2 \left(\frac{1}{x} \right)^6.$$

(18) Consider the motion of a vertical pendulum comprising a light stiff rod of length l with a mass m attached at one end. The other end is its point of support which oscillates horizontally, its distance from a fixed point 0 given by $a \sin \omega t$. If ω is large by comparison with the natural frequency of motion, find the mean motion Hamiltonian and its fixed points. Discuss the stability of these fixed points and determine the frequency of small oscillations about any that are stable.

(19) Consider a Hamiltonian of the form

$H(q, p, t) = H_0(q, p) + V_1(q) \sin \omega t + V_2(q) \sin 2\omega t$,

where $2\pi/\omega$ is small by comparison with the times during which the unperturbed system, $H_0(q, p)$, changes significantly. Show that the mean motion Hamiltonian is

$$K(\bar{q}, \bar{p}) = H_0(\bar{q}, \bar{p}) + \frac{1}{4} \frac{\partial^2 H_0}{\partial \bar{p}^2} \left[\frac{1}{\omega^2} \left(\frac{\partial V_1}{\partial \bar{q}} \right)^2 + \frac{1}{4\omega^2} \left(\frac{\partial V_2}{\partial \bar{q}} \right)^2 \right].$$

(20) Let $A(t)$ be a periodic function of time with period small by comparison with the times during which the system described by $H_0(q, p)$ changes significantly. Show that the mean motion Hamiltonian corresponding to

$$H(q, p, t) = H_0(q, p) + V(q) \frac{d^2 A}{dt^2}$$

is

$$K(q, p) = H_0(q, p) + \frac{1}{2} \left\langle \left(\frac{dA}{dt} \right)^2 \right\rangle \left(\frac{\partial V}{\partial q} \right)^2 \frac{\partial^2 H_0}{\partial p^2},$$

where $\langle \, \rangle$ denotes the mean over one period of $A(t)$.

(21) A ring of mass m slides smoothly on a rigid wire in the shape $z = f(x)$. The z-axis is vertically upwards and the x-axis is horizontal. If the wire is shaken vertically so that each point of the wire is displaced a vertical distance $\gamma(t)$, a periodic function of small period, show that the mean motion is described by the Hamiltonian

$$K(x, p) = \frac{p^2}{2m(1 + f'(x)^2)} + mgf(x) + \frac{1}{2} m \frac{\langle \dot{\gamma}^2 \rangle f'(x)^2}{1 + f'(x)^2}.$$

Discuss the motion in the following cases.

(a) In the neighbourhood of a local minimum of $f(x)$, where we can write

$$f(x) = \tfrac{1}{2} \alpha^2 x^2.$$

Show that the mean motion of small amplitude is simple harmonic and find the frequency.

(b) In the neighbourhood of a local maximum of $f(x)$, where

$$f(x) \simeq -\tfrac{1}{2} \alpha^2 x^2.$$

Find the condition that the mean motion can be simple harmonic and find the frequency in this case.

10 LINEAR SYSTEMS

10.1 Introduction

Autonomous linear systems were treated in the first four chapters. This chapter deals with non-autonomous linear systems for which the velocity function $v(\mathbf{r}, t)$ is an explicit function of time.

Section 10.2 contains the theory of first-order linear systems. It is very simple, but the methods can be extended to second-order systems, and an appreciation of their application to first-order systems makes the rest of the chapter much easier to understand. We restrict our attention to systems of two types.

The first type, considered in section 10.3, is an otherwise autonomous system subject to a driving force. These are the *forced* systems and the methods used are particularly important for linear electrical circuits.

In the second type, considered in the remainder of this chapter, the parameters used to define an autonomous system, such as the mass or length of a pendulum, or the capacitance in an electrical circuit, are made into functions of time. These are systems with *time-dependent parameters*. The theory of propagators is applied to these systems in section 10.4 and in 10.5 we restrict our attention to periodic variations of the parameters, leading to the important theory of linear maps and to questions of stability. The stability theory is worked out in detail for area-preserving linear maps in section 10.6, as it is required for applications to Hamiltonian systems. In section 10.7 the stability theory is applied to some problems of parametric resonance.

10.2 First-order systems

The general equation of motion (1.1) of a first-order system is linear when the velocity function $v(x, t)$ is linear in x, so that the general *linear* equation of motion has the form

$$\frac{\mathrm{d}x}{\mathrm{d}t} = v(x, t) = a(t)x + b(t). \tag{10.1}$$

If the system is forced then

$$a(t) = \text{constant} = a \tag{10.2}$$

and the equation of motion is

$$\frac{dx}{dt} - ax = b(t). \tag{10.3}$$

This equation is solved explicitly by multiplying by e^{-at} and then writing the resultant equation as

$$\frac{d}{dt}(x\,e^{-at}) = b(t)\,e^{-at}, \tag{10.4}$$

so that the solution with given $x(t_0)$ is

$$x(t) = e^{at}\left[x(t_0)\,e^{-at_0} + \int_{t_0}^{t} dt'\,b(t')\,e^{-at'} \right]. \tag{10.5}$$

If $a < 0$ then the system has a tendency to drift or '*decay*' exponentially towards the origin just like the unforced system of example 1.3 obtained by putting $b(t) = 0$. But, for the forced system, this tendency is counteracted by the forcing term whenever it acts.

Example 10.1

Obtain the motion of a system represented by a function $x(t)$ satisfying the differential equation

$$\frac{dx}{dt} + cx = 0 \quad (t < 0 \text{ and } t > 1) \tag{10.6}$$

$$= 1 \quad (0 \leqslant t \leqslant 1),$$

where $c > 0$, with the initial condition $x(-\infty) = 0$.

Clearly $x(t) = 0$ for $t \leqslant 0$. For $0 \leqslant t \leqslant 1$ we use equation (10.5) to give

$$x(t) = (1 - e^{-ct})/c \quad (0 \leqslant t \leqslant 1),$$

so that

$$x(1) = (1 - e^{-c})/c.$$

For $t > 1$ the system decays exponentially to give

$$x(t) = (e^{c} - 1)\,c^{-1}\,e^{-ct} \quad (t > 1)$$

or, graphically, as shown in the diagram.

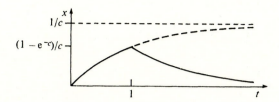

Now consider a first-order linear system with a time-dependent parameter. In this case $b(t) = 0$ and the equation of motion is

$$\frac{dx}{dt} = a(t)x. \tag{10.7}$$

This is linear and homogeneous in x, with an explicit solution given by

$$x(t) = x(t_0)\exp \int_{t_0}^{t} dt'\, a(t'). \tag{10.8}$$

We use the simplicity of these systems to introduce the theory of propagators, whose main application is to more complicated systems, some having no explicit solution.

It is clear from the differential equation (10.7) that the origin is a fixed point. Further, because the equation is linear and homogeneous in $x(t)$, it follows that if $x_1(t)$ is a solution, then so is $cx_1(t)$ for any constant c. Let $x_1(t)$ be the solution such that

$$x_1(t_0) = 1. \tag{10.9}$$

and let $x_1(t_1) = K$ for this solution. Then the solution $cx_1(t)$ has the value c at $t = t_0$ and the value cK at $t = t_1$, so for *all* solutions we have

$$x(t_1) = K\, x(t_0), \tag{10.10}$$

where the value of K depends upon t_0 and t_1, but not upon the choice of solution $x(t)$. Expressing this dependence explicitly we have, for all solutions of equation (10.7),

$$x(t_1) = K(t_1, t_0)x(t_0). \tag{10.11}$$

$K(t_1, t_0)$ is the *propagator* from time t_0 to time t_1 of the linear system whose motion is described by equation (10.7): note the order of the arguments. The theory is valid whether t_1 is later or earlier than t_0. Clearly when they are the same

$$K(t_0, t_0) = 1. \tag{10.12}$$

Given the state of the linear system at time t_0 the propagator $K(t_1, t_0)$ enables us to obtain the state at time t_1 without any further information. We represent this by diagram (10.13).

$$\xrightarrow{\hspace{1.5cm} K(t_1, t_0) \hspace{1.5cm}}$$

$$\begin{array}{lr} & (10.13) \\ t_0 \qquad\qquad\qquad t_1 & \end{array}$$

There are simple relations between propagators. Interchanging t_1 and t_0 in equation (10.11) and then using (10.11) again to eliminate $x(t_1)$ shows that, for any $x(t)$,

$$x(t_0) = K(t_0, t_1) x(t_1) = K(t_0, t_1) K(t_1, t_0) x(t_0), \qquad (10.14)$$

and so

$$K(t_0, t_1) K(t_1, t_0) = 1. \qquad (10.15a)$$

$$\begin{array}{lr} & (10.15b) \\ t_0 \qquad\qquad t_1 & \end{array}$$

The propagator from t_1 to t_0 is the inverse of the propagator from t_0 to t_1.

Substituting t_2 for t_1 and t_1 for t_0 in equation (10.11) gives

$$x(t_2) = K(t_2, t_1) x(t_1)$$

$$= K(t_2, t_1) K(t_1, t_0) x(t_0)$$

$$= K(t_2, t_0) x(t_0) \quad \text{(by definition)} \qquad (10.16)$$

and so the propagator from t_0 to t_2 is

$$K(t_2, t_0) = K(t_2, t_1) K(t_1, t_0). \qquad (10.17a)$$

$$\begin{array}{lr} & (10.17b) \\ t_0 \qquad t_1 \qquad t_2 & \end{array}$$

This is the multiplication rule for propagators. There is no need for $t_0 < t_1 < t_2$ as illustrated in the diagram. For example, if $t_1 < t_2 < t_0$ the diagram is as shown below.

$$\begin{array}{lr} & (10.18) \\ t_1 \qquad t_2 \qquad t_0 & \end{array}$$

Equation (10.17) is still valid. Notice that the first argument of the propagator function is always the time at the head of the corresponding arrow. It may be earlier or later than the second argument.

Example 10.2

The motion of a first-order linear system is represented by a function $x(t)$ satisfying equation (10.7). If $x(0) = 2$ then $x(1) = 10$ and if $x(1) = 3$ then $x(2) = 1$. Use the theory of propagators to determine $x(2)$ when $x(0) = -\frac{1}{5}$.

From the definition (10.11) of the propagator we have $K(1, 0) = 5$, $K(2, 1) = \frac{1}{3}$, so, by the multiplication rule, $K(2, 0) = \frac{5}{3}$ and, using (10.11) again, we have

$$x(2) = \left(\tfrac{5}{3}\right)\left(-\tfrac{1}{5}\right) = -\tfrac{1}{3} \quad \text{for } x(0) = -\tfrac{1}{5}.$$

For many systems it happens that $a(t) = \text{constant} = c$ in some interval of time from t_0 to t_1. In that case, the propagator from t_0 to t_1 is given by (10.8) and (10.11) as

$$K(t_1, t_0) = e^{c(t_1 - t_0)}. \tag{10.19}$$

Example 10.3

The motion of a system obeys the equations

$$\dot{x} = 0 \qquad (x < 0, x > 2)$$

$$= x \qquad (0 \leqslant x \leqslant 1)$$

$$= -2x \qquad (1 < x \leqslant 2).$$

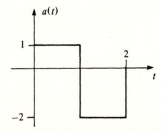

Find the propagators $K(3, -1)$ and $K(\frac{3}{2}, 0)$.

We split the time into segments of constant $a(t)$. For each of these segments, equation (10.19) provides the propagator, so that

$$K(0, -1) = 1, \quad K(1, 0) = e,$$

$$K(2, 1) = e^{-2}, K(3, 2) = 1.$$

Therefore $K(3, -1) = e^{-1}$. We also have $K(\frac{3}{2}, 1) = e^{-1}$, so

$$K(\tfrac{3}{2}, 0) = K(\tfrac{3}{2}, 1) K(1, 0) = 1.$$

Propagator theory is particularly valuable for linear homogeneous systems that satisfy $\dot{x} = a(t) x$, when $a(t)$ is a periodic function of time. These systems have periodic conditions, but their motion is not usually periodic.

Suppose $a(t)$ has period T, that is,

$$a(t + T) = a(t) \quad \text{(for all } t\text{)}. \tag{10.20}$$

Choose a convenient origin of time and define the *period propagator K* to be

$$K = K(T, 0). \tag{10.21}$$

Any $x(t)$ satisfying

$$\frac{dx(t)}{dt} = a(t) x(t) \tag{10.22}$$

also satisfies

$$\frac{dx(t + T)}{d(t + T)} = a(t + T) x(t + T). \tag{10.23}$$

Since T is constant and $a(t)$ is periodic, it also satisfies

$$\frac{dx(t + T)}{dt} = a(t) x(t + T), \tag{10.24}$$

so $x(t + T)$ is also a solution of the original equation of motion, and by induction, so is $x(t + nT)$ for any positive integer n. By a similar derivation it is also true for negative integers n.

Therefore, if for constant c

$$x(t_1) = c\, x(t_0),$$

then

$$x(t_1 + nT) = c\, x(t_0 + nT) \tag{10.25}$$

and, by the definition of the propagator,

$$K(t_1 + nT, t_0 + nT) = K(t_1, t_0) \quad (a(t) \text{ periodic with period } T). \tag{10.26}$$

In particular we have

$$K((n + 1) T, nT) = K \tag{10.27}$$

and, by the multiplication rule,

$$K(nT, 0) = K(nT, (n - 1) T)\, K((n - 1) T, (n - 2) T) \cdots K(T, 0)$$

$$= K^n \quad (n = 0, \pm 1, \pm 2 \ldots). \tag{10.28}$$

Thus, the effect of waiting for n periods is to raise the period propagator to the nth power.

$$K(nT, 0) = K^n \tag{10.29}$$

The period propagator K is all that we need in order to determine the behaviour of a periodic system at times nT. It provides the information needed to determine the states that would be 'seen' using a stroboscope which flashes on briefly at those times. These states define a sequence of points,

$$x_n = x(nT) = K^n x_0, \tag{10.30}$$

in the phase space. The origin is clearly a fixed point of the sequence, and its stability depends upon K. It follows from equation (10.8) that $K > 0$, but, for completeness in the next chapter, we also consider negative values of K.

If $|K| < 1$ then $\lim_{n \to \infty} x_n = 0$, but, if $|K| > 1$, then the sequence x_n diverges,

so the origin is either

a strongly stable fixed point if $|K| < 1$, $\tag{10.31a}$

or

an unstable fixed point if $|K| > 1$, $\tag{10.31b}$

of the sequence x_n defined by (10.30).

The various types of sequence look like this when $x_0 < 0$.

$$\tag{10.32}$$

When $K = \pm 1$ the origin is a fixed point and is stable but not strongly stable. An unstable fixed point about which the phase points oscillate (i.e. when $K < -1$) is also known as an *overstable* fixed point.

Example 10.4

A system with periodic conditions of period T satisfies the equation

$$\dot{x} = a_1 x \quad (0 \leqslant t < T_1)$$
$$= a_2 x \quad (T_1 \leqslant t < T). \tag{10.33}$$

For what values of a_1 and a_2 is the origin a stable fixed point for times nT with integer n?

The coefficient $a(t)$ is constant during the two intervals, so

$$K(T_1, 0) = e^{a_1 T_1}$$
$$K(T, T_1) = e^{a_2(T-T_1)}. \tag{10.34}$$

Therefore the period propagator is

$$K = K(T, 0) = \exp\left[a_1 T_1 + a_2(T - T_1)\right]. \tag{10.35}$$

The origin is a stable fixed point when

$$K < 1, \quad a_1 T_1 + a_2(T - T_1) < 0. \tag{10.36}$$

The state of the system as a function of time can be graphed for various values of the propagators. For example, if $a_1 = -a_2 = \frac{1}{2}$, $T_1 = 1$ and $T = 3$ the graph is as shown below, where the multiples of the period T are denoted by crosses.

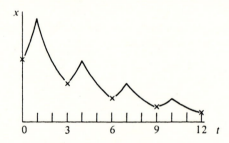

10.3 Forced linear oscillator

This Hamiltonian system of one degree of freedom is a forced system as defined in the introduction to this chapter, and is treated like the forced first-order systems of section 10.2.

Suppose we apply the force $F(t)$ to the mechanical linear oscillator of example 4.2 with mass m and angular frequency ω. The potential producing the force $F(t)$ is

$$V(q, t) = -q F(t), \tag{10.37}$$

so that the total Hamiltonian is

$$H(q, p, t) = p^2/2m + \tfrac{1}{2} m\omega^2 q^2 - q\, F(t) \tag{10.38}$$

and Hamilton's equations are

$$\dot{q} = p/m,$$

$$\dot{p} = -m\omega^2 q + F(t), \tag{10.39}$$

giving the second-order differential equation

$$\ddot{q} + \omega^2 q = F(t)/m. \tag{10.40}$$

This equation can be solved by first writing it as a differential equation for the complex variable

$$z = \dot{q} + i\omega q, \tag{10.41}$$

where

$$p = m\,\mathrm{Re}(z), \quad q = \mathrm{Im}(z)/\omega.$$

Then $\dot{z} = \ddot{q} + i\omega\dot{q}$ and, using equation (10.40), we have

$$\dot{z} - i\omega z = F(t)/m. \tag{10.42}$$

This has exactly the same form as equation (10.3) and can be solved by the same method to give

$$z(t) = e^{i\omega t} \left[z(t_0) e^{-i\omega t_0} + \frac{1}{m} \int_{t_0}^{t} \mathrm{d}t'\, F(t')\, e^{-i\omega t'} \right]. \tag{10.43}$$

The coordinate and momentum at time t are then given in terms of $q(t_0)$ and $p(t_0)$ by

$$q(t) = \mathrm{Im}(z(t))/\omega$$

$$= q(t_0) \cos \omega(t - t_0) + (m\omega)^{-1} p(t_0) \sin \omega(t - t_0)$$

$$+ (m\omega)^{-1} \int_{t_0}^{t} \mathrm{d}t'\, F(t') \sin \omega(t - t') \tag{10.44a}$$

$$p(t) = m\,\mathrm{Re}(z(t))$$

$$= p(t_0) \cos \omega(t - t_0) - m\omega\, q(t_0) \sin \omega(t - t_0)$$

$$+ \int_{t_0}^{t} \mathrm{d}t'\, F(t') \cos \omega(t - t'). \tag{10.44b}$$

The first two terms in each equation give the explicit solution for the unforced motion. Because the system is linear, the forcing terms are simply added to these.

The energy E of a forced system is not conserved when defined, as it is, without the applied force $F(t)$. If the system starts from rest at the origin, then, for $t > t_0$,

$$E(t) = p^2/2m + \tfrac{1}{2} m\omega^2 q^2 = \tfrac{1}{2} m \ |z(t)|^2$$

$$= \frac{1}{2m} \left| \int_{t_0}^{t} dt' \, F(t') \, e^{-i\omega t'} \right|^2 \tag{10.45}$$

Example 10.5

Suppose a forced linear oscillator with coordinate q satisfying equation (10.40) is at rest at the origin before $t = 0$, i.e.

$$q = \dot{q} = F(t) = 0 \quad (t < 0) \tag{10.46}$$

and that afterwards the applied force is

$$F(t) = F_0 \cos \Omega t = ma \cos \Omega t \quad (t \geqslant 0), \tag{10.47}$$

where $\Omega > 0$, $\Omega \neq \omega$ and ω is the angular frequency of the oscillator.

By equation (10.43), for $t \geqslant 0$ and $z = \dot{q} + i\omega q$,

$$z(t) = a \, e^{i\omega t} \int_{0}^{t} dt' \cos \Omega t' \, e^{-i\omega t'}$$

$$= a \, e^{i\omega t} \frac{1}{2} \left[\frac{e^{i(\Omega - \omega)t'}}{i(\Omega - \omega)} + \frac{e^{-i(\Omega + \omega)t'}}{-i(\Omega + \omega)} \right]_{0}^{t}$$

$$= \frac{a}{2i} \left[\frac{e^{i\Omega t} - e^{i\omega t}}{\Omega - \omega} - \frac{e^{-i\Omega t} - e^{i\omega t}}{\Omega + \omega} \right] \tag{10.48}$$

$$q(t) = \mathrm{Im}(z)/\omega$$

$$= \frac{-a}{2\omega} \left[\frac{\cos \Omega t - \cos \omega t}{\Omega - \omega} - \frac{\cos \Omega t - \cos \omega t}{\Omega + \omega} \right]$$

$$= a \left[\frac{\cos \omega t - \cos \Omega t}{\Omega^2 - \omega^2} \right]. \tag{10.49}$$

After $t = 0$ the displacement $q(t)$ has components with both the natural frequency ω and the 'forcing' frequency Ω. The possible displacement cannot exceed

$$q_{\max} = \frac{2a}{\Omega^2 - \omega^2} . \tag{10.50}$$

For fixed ω this increases without limit as the forcing frequency Ω approaches the natural frequency ω, that is, as the applied force approaches the *resonance condition* where $\Omega = \omega$.

The phenomenon of resonance is commonly observed. Ships and boats pitch and roll much more when pushed by waves near a resonant frequency. Piano and guitar strings vibrate when tuned to the same frequency as the voice of a nearby singer. Parts of cars, trains and planes start to vibrate when the vehicles are moving at particular speeds for similar reasons.

In practice for real systems damping and non-linear effects (see e.g. exercise 7.15) that are not included in this example always appear for sufficiently large amplitudes of the motion and the motion remains bounded even when $\Omega = \omega$.

The complex variable method of this section can be extended to much more complicated linear systems, with very wide application, particularly to electrical circuits.

10.4   Propagators

Propagators of first-order linear systems were introduced in section 10.2. Their theory extends naturally to systems of second- and higher-order, using matrices. The main additional features are the commutation properties of the matrices and the greater variety in the types of stability for periodic velocity functions $v(\mathbf{r}, t)$.

The extension of the second-order theory to higher orders is obvious and is omitted as usual.

When a second-order system is linear, the velocity function $v(\mathbf{r}, t)$ is linear in the state vector $\mathbf{r} = (x, y)$ so that the equations of motion (3.1) can be written

$$\frac{d\mathbf{r}}{dt} = v(\mathbf{r}, t) = A(t)\,\mathbf{r}(t) + \mathbf{b}(t), \tag{10.51}$$

where $A(t)$ is a real 2×2 matrix function of t and $\mathbf{b}(t)$ is a real column 2-vector function of t. If the system is of the second type then $\mathbf{b}(t) = 0$ and the equation of motion becomes

$$\frac{d\mathbf{r}}{dt} = A(t)\mathbf{r} \tag{10.52}$$

or

$$
\left.
\begin{aligned}
\frac{dx}{dt} &= A_{11}(t)x + A_{12}(t)y \\[2mm]
\frac{dy}{dt} &= A_{21}(t)x + A_{22}(t)y.
\end{aligned}
\right\} \tag{10.53}
$$

The equation (10.52) is linear and homogeneous in the column vector state variable $\mathbf{r}$, so if $\mathbf{r}_1(t)$ and $\mathbf{r}_2(t)$ are any pair of solutions, then for any constants c_1 and c_2, the linear combination

$$\mathbf{r}(t) = c_1 \mathbf{r}_1(t) + c_2 \mathbf{r}_2(t) \tag{10.54}$$

is also a solution.

Let $\mathbf{r}_1(t)$ and $\mathbf{r}_2(t)$ be the special solutions such that, for a given t_0,

$$\mathbf{r}_1(t_0) = \begin{pmatrix} x_1(t_0) \\ y_1(t_0) \end{pmatrix} = \begin{pmatrix} 1 \\ 0 \end{pmatrix} \tag{10.55a}$$

$$\mathbf{r}_2(t_0) = \begin{pmatrix} x_2(t_0) \\ y_2(t_0) \end{pmatrix} = \begin{pmatrix} 0 \\ 1 \end{pmatrix} \tag{10.55b}$$

and let us denote the values of these solutions at time t_1 by K_{ij}, where

$$\mathbf{r}_j(t_1) = \begin{pmatrix} x_j(t_1) \\ y_j(t_1) \end{pmatrix} = \begin{pmatrix} K_{1j} \\ K_{2j} \end{pmatrix} \quad (j = 1, 2). \tag{10.56}$$

On substituting $t = t_0$ in equation (10.54) the values of c_1 and c_2 are found to be

$$c_1 = x(t_0), \quad c_2 = y(t_0). \tag{10.57}$$

On substituting $t = t_1$ in equation (10.54) and using (10.56) we find that

$$\begin{aligned}
\begin{pmatrix} x(t_1) \\ y(t_1) \end{pmatrix} &= c_1 \begin{pmatrix} x_1(t_1) \\ y_1(t_1) \end{pmatrix} + c_2 \begin{pmatrix} x_2(t_1) \\ y_2(t_1) \end{pmatrix} \\
&= c_1 \begin{pmatrix} K_{11} \\ K_{21} \end{pmatrix} + c_2 \begin{pmatrix} K_{12} \\ K_{22} \end{pmatrix} \\
&= \begin{pmatrix} K_{11} & K_{12} \\ K_{21} & K_{22} \end{pmatrix} \begin{pmatrix} x(t_0) \\ y(t_0) \end{pmatrix}
\end{aligned} \tag{10.58}$$

and this is written as

$$\mathbf{r}(t_1) = K\,\mathbf{r}(t_0), \tag{10.59}$$

where

$$K = \begin{pmatrix} K_{11} & K_{12} \\ K_{21} & K_{22} \end{pmatrix} \tag{10.60}$$

is a matrix whose value depends upon t_0 and t_1. Since any solution $\mathbf{r}(t)$ for the second-order system can be written in the form (10.54), K does not depend upon the choice of solutions $\mathbf{r}(t)$. Expressing the dependence on t_1 and t_0 explicitly we write, for any $\mathbf{r}(t)$,

$$\mathbf{r}(t_1) = K(t_1, t_0)\,\mathbf{r}(t_0). \tag{10.61}$$

Each of the four elements of the 2×2 matrix $K(t_1, t_0)$ depends on t_1 and t_0.

The matrix $K(t_1, t_0)$ is the *propagator* from time t_0 to time t_1. It defines the linear transformation from the points in the phase plane at time t_0 to the points at time t_1.

If $t_1 = t_0$ the system remains unchanged and K is the unit matrix

$$K(t_0, t_0) = I. \tag{10.62}$$

Interchanging t_1 and t_0 in equation (10.61) and then using (10.61) to eliminate $\mathbf{r}(t_1)$ we find that for *any* $\mathbf{r}(t)$ satisfying (10.52)

$$\mathbf{r}(t_0) = K(t_0, t_1) \mathbf{r}(t_1)$$
$$= K(t_0, t_1) K(t_1, t_0) \mathbf{r}(t_0) \tag{10.63}$$

so that

$$K(t_0, t_1) K(t_1, t_0) = I. \tag{10.64}$$

This is similar to the first-order relation (10.15). The backward propagator is the inverse matrix of the forward propagator.

Substituting t_0 for t_1 and t_2 for t_1 in equation (10.61), we have, for any $\mathbf{r}(t)$ satisfying (10.52),

$$\mathbf{r}(t_2) = K(t_2, t_1) \mathbf{r}(t_1)$$
$$= K(t_2, t_1) K(t_1, t_0) \mathbf{r}(t_0)$$
$$= K(t_2, t_0) \mathbf{r}(t_0) \quad \text{(by definition)} \tag{10.65}$$

and so the propagator from t_0 to t_2 is

$$K(t_2, t_0) = K(t_2, t_1) K(t_1, t_0). \tag{10.66}$$

This is the multiplication rule for second-order propagators. Unlike the first-order relation (10.17) the product cannot be written the other way round because the two matrices may not commute. In general the second-order relations resemble the first-order relations, except that the commutation properties must be carefully watched.

Example 10.6

Find the propagators $K(t_1, t_0)$ at arbitrary times t_0 and t_1 for the linear oscillator and linear repulsive force of examples 4.2 and 4.3.

As usual the coordinate and momentum (q, p) take the place of (x, y). For the linear oscillator with mass m and potential $V(q) = \frac{1}{2} a q^2$, $a > 0$, the equation of motion is $\ddot{q} = -\omega^2 q$, with $\omega = (a/m)^{\frac{1}{2}}$. The general solution for the state at time t in terms of the state at time t_0 is given by equations (10.44) with $F = 0$. The special solutions needed for the propagator are $\mathbf{r}_1(t)$ with $q_1(t_0) = 1$, $p_1(t_0) = 0$, giving

$$\mathbf{r}_1(t_1) = \begin{pmatrix} q_1(t_1) \\ p_1(t_1) \end{pmatrix} = \begin{pmatrix} K_{11}(t_1, t_0) \\ K_{21}(t_1, t_0) \end{pmatrix}$$

$$= \begin{pmatrix} \cos \omega(t_1 - t_0) \\ -m\omega \sin \omega(t_1 - t_0) \end{pmatrix} \tag{10.67}$$

and $\mathbf{r}_2(t)$ with $q_2(t_0) = 0$, $p_2(t_0) = 1$, giving

$$\mathbf{r}_2(t_1) = \begin{pmatrix} q_2(t_1) \\ p_2(t_1) \end{pmatrix} = \begin{pmatrix} K_{12}(t_1, t_0) \\ K_{22}(t_1, t_0) \end{pmatrix}$$

$$= \begin{pmatrix} (m\omega)^{-1} \sin \omega(t_1 - t_0) \\ \cos \omega(t_1 - t_0) \end{pmatrix}. \tag{10.68}$$

The propagator for the linear oscillator (LO) is, therefore,

$$K_{LO}(t_1, t_0) = \begin{pmatrix} \cos \omega(t_1 - t_0) & (m\omega)^{-1} \sin \omega(t_1 - t_0) \\ -m\omega \sin \omega(t_1 - t_0) & \cos \omega(t_1 - t_0) \end{pmatrix}. \tag{10.69}$$

For $m\omega = 1$ this represents a finite rotation through an angle $-\omega(t_1 - t_0)$ about the origin in the phase plane. For $t_1 > t_0$ the rotation is clockwise, in a *negative* sense according to the standard convention, as expected from diagram (4.20).

For $m\omega \neq 1$ the propagator defines a motion around an ellipse of constant energy:

$$E = p^2/2m + \tfrac{1}{2} m\omega^2 q^2 \tag{10.70a}$$

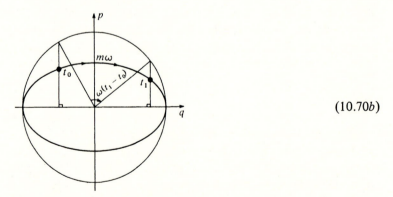

$$\tag{10.70b}$$

The state $\mathbf{r}$ at time t_1 is found by projecting vertically from the point $\mathbf{r}(t_0)$ on the ellipse to a point on the circle, rotating through the angle $| \omega(t_1 - t_0) |$, and projecting vertically back onto the ellipse.

For the linear repulsive force $a < 0$ and we put $\gamma = (-a/m)^{\frac{1}{2}}$. The equation of motion $\ddot{q} = \gamma^2 q$ has the general solution

$$q = A_+ e^{\gamma t} + A_- e^{-\gamma t},$$

$$p = m\gamma(A_+ e^{\gamma t} - A_- e^{-\gamma t}). \tag{10.71}$$

The special solution $r_1(t)$ of (10.55a) has initial conditions $A_+ e^{\gamma t_0} = A_- e^{-\gamma t_0} = \frac{1}{2}$ so that $K_{11} = \cosh \gamma(t - t_0)$, $K_{21} = m\gamma \sinh \gamma(t - t_0)$. Also the special solution $r_2(t)$ of (10.55b) has initial conditions $A_+ e^{\gamma t_0} = -A_- e^{-\gamma t_0} = (2m\gamma)^{-1}$ so that $K_{12} = (m\gamma)^{-1} \sinh \gamma(t - t_0)$, $K_{22} = \cosh \gamma(t - t_0)$. The propagator is

$$K(t_1, t_0) = \begin{pmatrix} \cosh \gamma(t_1 - t_0) & (m\gamma)^{-1} \sinh \gamma(t_1 - t_0) \\ m\gamma \sinh \gamma(t_1 - t_0) & \cosh \gamma(t_1 - t_0) \end{pmatrix}. \tag{10.72}$$

This represents a motion along one branch of the hyperbola of constant energy

$$E = p^2/2m - \tfrac{1}{2} m\gamma^2 q^2. \tag{10.73a}$$

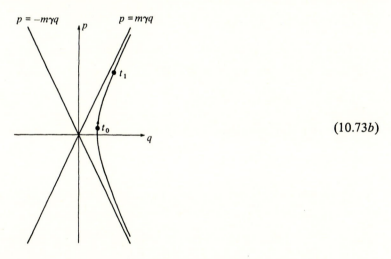

$$\tag{10.73b}$$

The propagators (10.69) and (10.72) both produce linear area-preserving transformations of the phase plane.

Example 10.7

Find the propagator $K_0(t_1, t_0)$ for a free particle of mass m, and the propagator $K_1(t_1, t_0)$ for a particle that moves freely except for an impulse

$$P(q) = aq \tag{10.74}$$

acting at a time t'. Assume $t_1 \geqslant t' \geqslant t_0$.

Given $q(t_0)$ and $p(t_0)$ the coordinate $q(t)$ and momentum $p(t)$ of a free particle are

$$q(t) = q(t_0) + (t - t_0) p(t_0)/m,$$

(10.75)

$$p(t) = p(t_0),$$

so the free particle propagator is

$$K_0(t_1, t_0) = \begin{pmatrix} 1 & (t_1 - t_0)/m \\ 0 & 1 \end{pmatrix}.$$

(10.76)

An impulse is a very large force acting for a very short time, resulting in a significant change P in the momentum p, but negligible change in the coordinate q. Impulses may be the result of collisions or rapidly changing fields and if the field depends upon q, then so does the impulse, as in our case.

From the definition of an impulse, at time t' in our case

$$\lim_{\epsilon \to 0+} q(t' + \epsilon) = \lim_{\epsilon \to 0+} q(t' - \epsilon)$$

(10.77a)

$$\lim_{\epsilon \to 0+} p(t' + \epsilon) = \lim_{\epsilon \to 0+} p(t' - \epsilon) + aq(t'),$$

(10.77b)

where the limit $\epsilon \to 0+$ is taken through positive values of ϵ. Thus the effect of the impulse can be described by a propagator

$$K_I = \lim_{\epsilon \to 0+} K(t' + \epsilon, t' - \epsilon) = \begin{pmatrix} 1 & 0 \\ a & 1 \end{pmatrix}.$$

(10.78)

If $t_1 > t' > t_0$ then, by the product rule,

$$K_1(t_1, t_0) = K_0(t_1, t') K_I K_0(t', t_0)$$

(10.79)

$$= \begin{pmatrix} 1 + a(t_1 - t')/m & (t_1 - t_0)/m + a(t_1 - t')(t' - t_0)/m^2 \\ a & 1 + a(t' - t_0)/m \end{pmatrix}.$$

(10.80)

We have not yet considered every case, as either t_0 or t_1 may be equal to t'. But then there is an ambiguity because the limits from positive and negative times are different. It is very useful to adopt a definite convention.

According to the *impulse convention*, the state of a system at a time t' when an impulse is applied is the limiting state just *before* the time t'.

$$q(t') = \lim_{\epsilon \to 0+} q(t' - \epsilon),$$

$$p(t') = \lim_{\epsilon \to 0+} p(t' - \epsilon).$$ (10.81)

With this convention the propagators are

$$K_1(t', t_0) = \lim_{\epsilon \to 0+} K_1(t' - \epsilon, t_0)$$

$$= K_0(t', t_0)$$ (10.82)

$$K_1(t_1, t') = \lim_{\epsilon \to 0+} K_1(t_1, t' - \epsilon)$$

$$= \lim_{\epsilon \to 0+} K_0(t_1, t') \, K_I \, K(t', t' - \epsilon)$$ (10.83)

$$= \begin{pmatrix} 1 + a(t_1 - t')/m & (t_1 - t')/m \\ a & 1 \end{pmatrix}.$$ (10.84)

10.5   Periodic conditions and linear maps

Now consider second-order systems with periodic conditions. These are very common and important in astronomy, electrical circuit theory and solid state physics. Furthermore, the theory is essential to the non-linear theory of periodic Hamiltonians presented in chapter 11. However there is a close parallel between the second- and first-order theories so the reader should refer back to section 10.1, starting just before equation (10.20).

A linear system with periodic time-dependent parameters satisfies equation (10.52), $\dot{\mathbf{r}} = A(t)\,\mathbf{r}$, where the matrix $A(t)$ is periodic, of period T,

$$A(t + T) = A(t) \quad \text{(for all } t\text{).}$$ (10.85)

Choose a convenient origin of time and define the period propagator K to be

$$K = K(T, 0).$$ (10.86)

In the same way as for first-order systems, for which the proof has been given in section 10.2, if $\mathbf{r}(t)$ satisfies equation (10.52) with the periodicity condition (10.85), then so does $\mathbf{r}(t + nT)$ for any integer n, positive or negative.

But the solution $\mathbf{r}_1(t)$ of equation (10.52) with initial condition (10.55a) has the values

$$\begin{pmatrix} 1 \\ 0 \end{pmatrix} \text{ at } t_0 \quad \text{and} \quad \begin{pmatrix} K_{11} \\ K_{21} \end{pmatrix} \text{ at } t_1,$$ (10.87)

so $\mathbf{r}_1(t + nT)$ satisfies (10.52) and has the values

$$\begin{pmatrix} 1 \\ 0 \end{pmatrix} \quad \text{at} \quad t_0 + nT \quad \text{and} \quad \begin{pmatrix} K_{11} \\ K_{21} \end{pmatrix} \quad \text{at} \quad t_1 + nT. \tag{10.88}$$

Therefore $K_{11}(t_1 + nT, t_0 + nT) = K_{11}(t_1, t_0)$ and $K_{21}(t_1 + nT, t_0 + nT) = K_{21}(t_1, t_0)$, and a similar argument can be used for K_{12} and K_{22}. Consequently we have

$$K(t_1 + nT, t_0 + nT) = K(t_1, t_0) \qquad (A(t + T) = A(t)) \tag{10.89}$$

and, in particular,

$$K((n + 1)T, nT) = K, \tag{10.90}$$

so that, by the multiplication rule,

$$K(nT, 0) = K^n. \tag{10.91}$$

As for first-order systems, the period propagator K is all that is needed to determine the behaviour of the system at times nT.

The period propagator K defines a linear transformation or *linear map* from the phase space onto itself:

$$\mathbf{r}' = K\mathbf{r}. \tag{10.92a}$$

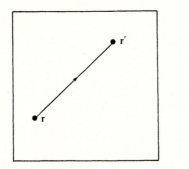

$$\tag{10.92b}$$

Using the notation

$$\mathbf{r}(nT) = \mathbf{r}_n, \tag{10.93}$$

equation (10.92) relates successive $\mathbf{r}_n$ by a single operation of the linear map

$$\mathbf{r}_{n + 1} = K\mathbf{r}_n, \tag{10.94}$$

whereas equation (10.91) gives $\mathbf{r}_n$ in terms of $\mathbf{r}_0$ by n operations of the map,

$$\mathbf{r}_n = K^n \mathbf{r}_0. \tag{10.95}$$

The map defined by K is said to generate the sequence of points $\mathbf{r}_n = \mathbf{r}(nT)$ defining the states at times $t = nT$ of the systems whose motion satisfies the differential equation (10.52) with the periodicity condition (10.85).

The most important properties of the periodic system can be derived from the study of repeated applications of the linear map and we concentrate on these because they are simpler, leaving to one side all other properties of the periodic system. Sometimes there is no choice, because the state of the system can only be known at the times nT. For example, the measurement of an animal population may be made annually with $T = 1$ year. The population at intermediate times may not be known with any precision, but it may be possible to estimate the propagator K and thus the long time behaviour of the population at times nT. This can be valuable when the result concerns the extinction of a species.

10.6   Linear area-preserving maps

The theory of linear maps and the development of a determinate system at discrete intervals of time, as given by the evolution equation (10.94), is similar to the theory of linear autonomous systems of chapter 3, where continuous dependence on time is given by the evolution equation (3.38). Both depend upon the properties of 2×2 matrices given in chapter 2 and the stability theories have much in common. For simplicity we do not go into detail on all the different types of stability of linear maps but consider only the area-preserving maps as required for Hamiltonian systems. For such maps the period propagator has unit determinant,

$$\det K = 1. \tag{10.96}$$

We study the linear transformation generated by the matrix K with the phase space coordinate vector $\mathbf{r}$ by finding a new coordinate

$$\mathbf{R} = \begin{pmatrix} X \\ Y \end{pmatrix} \tag{10.97}$$

obtained from $\mathbf{r}$ by a particular linear transformation

$$\mathbf{R} = M\mathbf{r}, \tag{10.98}$$

where M is a non-singular transformation matrix. In the new coordinate system the map is represented by the equation

$$\mathbf{R}' = L\mathbf{R}, \quad \text{where} \quad L = MKM^{-1}. \tag{10.99}$$

and, because K is area-preserving, so is L:

$$\det L = 1. \tag{10.100}$$

According to the general theory of linear transformations in chapter 2, for any K a non-singular 2×2 matrix M can be found so that L is of type 1, 2 or 3 as defined in chapter 2, and each of these can be analysed in detail as in chapter 3.

The stability properties of the linear map depend on the eigenvalues k_1 and k_2 of the propagator K which are the diagonal elements of L. Particularly important is the trace

$$\tau = K_{11} + K_{22} = k_1 + k_2, \tag{10.101}$$

as shown in chapter 2. We confine our attention to linear maps with hyperbolic fixed points, when $|\tau| > 2$, and with elliptic fixed points, when $|\tau| < 2$. These are named *hyperbolic* and *elliptic maps* and, according to the definitions of chapter 2, they are of type 1 and 2 respectively.

Hyperbolic maps: we have $|\tau| > 2$ and the eigenvalues are real and distinct. L has the form

$$L = \begin{pmatrix} k_1 & 0 \\ 0 & k_2 \end{pmatrix} \tag{10.102}$$

and, because its determinant is unity, $k_2 = k_1^{-1}$ and

$$\text{either } k_1, k_2 > 0 \quad (type\ 1^+) \tag{10.103a}$$

$$\text{or} \quad k_1, k_2 < 0, (type\ 1^-) \tag{10.103b}$$

defining types 1^+ and 1^-.

From equations (10.99) and (10.102) the coordinates X_n, Y_n of $\mathbf{R}_n$ are decoupled and satisfy the equations

$$X_{n+1} = k_1 X_n,$$
$$Y_{n+1} = k_2 Y_n = Y_n/k_1 \tag{10.104}$$

so the product $X_n Y_n$ remains constant.

The orbit $\mathbf{R}_n$ lies on a rectangular hyperbola with X and Y axes as asymptotes as illustrated in figure 10.1, where the letters A, B, C, D denote successive quadrants of the $\mathbf{R}$-plane.

For type 1^+ either $k_1 > 1$ or $k_2 > 1$. If $k_1 > 1$ then $0_A, 1_A, 2_A, 3_A, 4_A$ are successive phase points of a possible motion, as are $0_B, 1_B, 2_B, 3_B, 4_B$ etc. If $k_1 < 1$ then $k_2 > 1$ and the same sequences appear in the reverse order.

For type 1^-, either $|k_1| > 1$, or $|k_2| > 1$. If $|k_1| > 1$ then $0_A, 1_C, 2_A, 3_C, 4_A$ are successive phase points of a possible motion, as are $0_B, 1_D, 2_B, 3_D, 4_B$ and similarly $0_C, 1_A, 2_C, 3_A, 4_C$ and $0_D, 1_B, 2_D, 3_B, 4_D$.

For $|k_1| < 1$ the same sequences appear in the reverse order. The motion switches between branches of the same hyperbola.

Fig. 10.1 Motion in the XY-plane for a hyperbolic map.

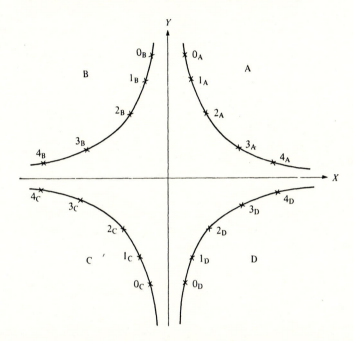

When the K matrix is of type 1, the origin is an *unstable hyperbolic* point of the linear map generated by K. We use the same description for the matrix and the linear map. The motion for type 1^- has an oscillatory component and is sometimes named *overstable* motion as it becomes unstable by swinging beyond the equilibrium point.

In the original **r** representation the orbits lie on hyperbolas whose asymptotes may be any two distinct straight lines through the origin.

Elliptic maps: we have $|\tau| < 2$ and the eigenvalues are complex. From chapter 2, there is a real angle α such that

$$k_1 = e^{i\alpha},$$

$$k_2 = e^{-i\alpha},$$

$$2\cos\alpha = \tau = K_{11} + K_{22}. \tag{10.105}$$

The K matrix is transformed to the L matrix,

$$L(\alpha) = \begin{pmatrix} \cos\alpha & -\sin\alpha \\ \sin\alpha & \cos\alpha \end{pmatrix} \tag{10.106}$$

and the linear transformation (10.99) produced by $L(\alpha)$ is a rotation through an angle α about the origin of the space with X and Y as Cartesian coordinates.

The successive points $\mathbf{R}_n$ of an orbit of the map are spaced at equal angles α around a circle.

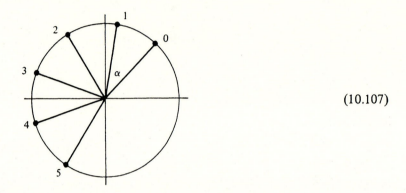

(10.107)

The nature of the orbit changes with arbitrarily small changes in the angle α. For example suppose that the fundamental period is T and

$$\alpha/2\pi = \tfrac{1}{5}.$$

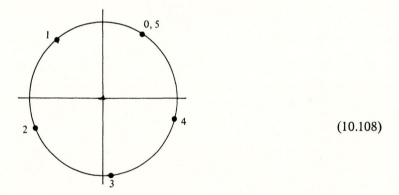

(10.108)

Then the system returns to its initial state after a period $5T$ and the motion passes through only 5 points. Similarly, if

$$\alpha/2\pi = \tfrac{2}{5}$$

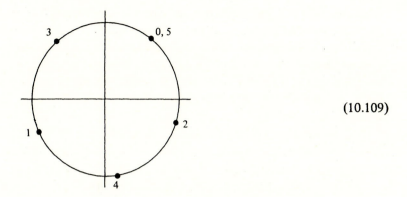

(10.109)

the system will also return after a period $5T$ after winding around the circle twice.

In general, suppose $\alpha/2\pi$, the *winding number*, is rational,

$$\alpha/2\pi = N/D \quad (N = 0, \pm1, \pm2, \ldots, D = 1, 2, \ldots), \tag{10.110}$$

where, by convention, the numerator N and denominator D have no common divisors. Then the orbits are periodic of period DT and wind around the circle N times in that period. Each orbit consists of just D points.

The Dth power of the propagator represents a rotation through $2N\pi$, which is the identity, so

$$K^D = 1 \quad (\alpha/2\pi = N/D). \tag{10.111}$$

The periodicity can also be expressed in terms of either the new or the original coordinate vectors,

$$\mathbf{R}_{n+D} = \mathbf{R}_n \quad (n = 0, \pm1, \pm2 \ldots) \tag{10.112}$$

$$\mathbf{r}_{n+D} = \mathbf{r}_n \quad (n = 0, \pm1, \pm2 \ldots). \tag{10.113}$$

In the original $\mathbf{r} = (x, y)$ representation, the points $\mathbf{r}_n$ lie on an ellipse. Thus, when the K matrix is of type 2 with $|\tau| < 2$, the origin is a *stable elliptic* point of the map generated by K and the same name is applied to K and to the map.

Now suppose the winding number, $\alpha/2\pi$, is irrational. Then the system does not return to its initial state after any multiple of the time T and the motion is not periodic. After a sufficient length of time the points $\mathbf{R}_n$ approach arbitrarily close to any point of the circle $|\mathbf{R}| = |\mathbf{R}_0|$ in the XY-plane and the points $\mathbf{r}_n$ approach arbitrarily close to any point on the corresponding ellipse. The origin is still a stable elliptic point.

Because there are rational numbers arbitrarily close to any irrational number there are periodic orbits for values of $\alpha/2\pi$ arbitrarily close to any irrational value and the qualitative behaviour of the system changes for arbitrarily small changes in α. The system is structurally unstable for changes in α.

Of course, in a real system there are all kinds of small perturbations, including terms that make the system non-Hamiltonian or non-linear, and the system cannot be followed for arbitrarily long periods of time, so that the distinction between two very close values of α is difficult to observe in practice. Nevertheless this structural instability makes considerable difficulties for the general theory of non-linear systems and gives rise to important new phenomena, as shown in chapter 11.

We have now obtained many properties of linear systems with periodic conditions from the period propagator K whose definition (10.86) depends on the choice of a convenient origin of time. In exercise 10.19 the reader is invited to show that these properties are independent of this choice of origin.

We have confined our attention to the most important examples of fixed points of linear area-preserving maps. There are many other types of fixed points of linear maps corresponding to the stable and unstable proper and improper nodes, spiral points and stars of chapter 3 and also the hyperbolic and elliptic points of maps that do not preserve area.

10.7 Periodic forces and parametric resonance

Now we apply the theory of the last three sections to systems with forces proportional to the displacement q and periodic in time. The simplest examples include impulses and we now consider them.

Example 10.8

Obtain the period propagator and investigate the stability of a particle of mass m and coordinate q subject to an infinite sequence of impulses

$$P(q) = aq \quad \text{at times} \quad t = nT \quad (n = 0, \pm 1, \pm 2 \dots) \tag{10.114}$$

and also illustrate the motion in configuration and in phase space at times nT.

For $a > 0$ the impulse always pushes the particle away from the origin and we should expect instability. When $a < 0$ the impulse always pushes the particle towards the origin and we should expect stability unless it pushes too hard. These expectations are confirmed by the analysis.

The period propagator K is obtained directly from equation (10.84) of example 10.7. Only the impulse at $t' = 0$ need be considered, to give

$$K = K_1(T, 0) = \begin{pmatrix} 1 + aT/m & T/m \\ a & 1 \end{pmatrix} \tag{10.115}$$

and its trace is

$$\tau = 2 + aT/m. \tag{10.116}$$

From the results of section 10.6 the stability properties are:

unstable for $aT/m > 0$;   (10.117a)

stable for $-4 < aT/m < 0$;   (10.117b)

unstable (overstable) for $aT/m < -4$.   (10.117c)

For our illustrations, suppose $T/m = 1$ throughout.
For the stable motion let $a = -1$; then

$$K = \begin{pmatrix} 0 & 1 \\ -1 & 1 \end{pmatrix}$$ (10.118)

and a possible motion is

$$\mathbf{r}_0 = \begin{pmatrix} 1 \\ 0 \end{pmatrix}, \quad \mathbf{r}_1 = \begin{pmatrix} 0 \\ -1 \end{pmatrix}, \quad \mathbf{r}_2 = \begin{pmatrix} -1 \\ -1 \end{pmatrix}, \quad \mathbf{r}_3 = \begin{pmatrix} -1 \\ 0 \end{pmatrix},$$

$$\mathbf{r}_4 = \begin{pmatrix} 0 \\ 1 \end{pmatrix}, \quad \mathbf{r}_5 = \begin{pmatrix} 1 \\ 1 \end{pmatrix}, \quad \mathbf{r}_6 = \begin{pmatrix} 1 \\ 0 \end{pmatrix}$$ (10.119)

and continuing cyclically. All orbits of the map with this propagator are periodic of period $6T$.

(10.120a)   (10.120b)

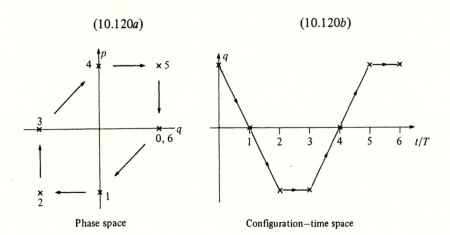

Phase space   Configuration–time space

The impulses push the particle back towards the origin. Because no forces are acting between the times nT, the complete motion at all times can be illustrated in this example by joining the points $\mathbf{r}_n$ by straight lines. By the theory of section 10.6 a new representation $(X, Y) = \mathbf{R}$ can be found in which the propagator is represented by a rotation matrix $L(-\alpha)$ like (10.106) with

$$2 \cos \alpha = \operatorname{Tr} K = 2 + aT/m \quad (\alpha > 0). \tag{10.121}$$

Note that the rotation is clockwise. This is typical of Newtonian systems. In our example $\cos \alpha = \frac{1}{2}$ so the rotation is through $60°$ and in this new representation the successive $\mathbf{R}_n$ are at the vertices of a hexagon. For arbitrary a in the range (10.117b), all the complications discussed at the end of the previous section occur. Normally the motion is not periodic and there are an infinite number of points on the corresponding ellipse.

For unstable motion of type 1^+ with $\tau > 2$, let $a = 1$ with the same initial conditions. The impulses all push the particle away from the origin,

$$K = \begin{pmatrix} 2 & 1 \\ 1 & 1 \end{pmatrix}$$

and the diagrams are as shown below.

(10.122a)   (10.122b)

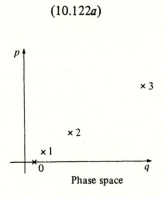

Phase space

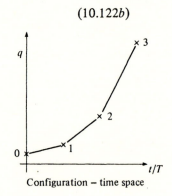

Configuration – time space

For unstable or overstable motion $\tau < -2$, type 1^-, let $a = -5$. The impulses push the particle very hard beyond the origin,

$$K = \begin{pmatrix} -4 & 1 \\ -5 & 1 \end{pmatrix}$$

and the diagrams are the ones below, where the oscillatory behaviour is clear.

(10.123*a*)   (10.123*b*)

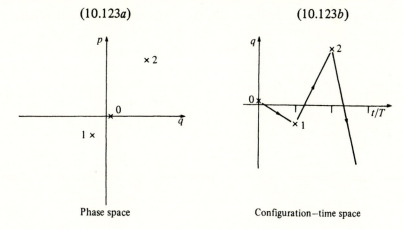

Phase space   Configuration–time space

The next example is one of an important class in which an otherwise autonomous linear Hamiltonian system is subject to periodic changes in its parameters, like a child on a swing who changes its effective length by altering the position of his body, or a microphone capacitor whose capacitance is changed periodically in response to a steady note from a musical instrument. Such parametric changes are of a different kind from those due to applied external forces, such as someone *pushing* a swing or *charging* a capacitor, for which the theory of section 10.3 is required.

Example 10.9

Investigate the stability of motion of a particle of mass m and coordinate q, subject both to a linear restoring force $-m\omega^2 q$ and to the infinite sequence of impulses (10.114) of the previous example 10.8. Assume that $a/m\omega$ is small and positive.

In the absence of the impulses we have a linear oscillator of angular frequency ω. Each impulse may be considered as a large and sudden, but temporary, change in the coefficient of q, so this is a parametric change. Between the times nT the propagator is that for the linear oscillator as given by K_{LO} in equation (10.69). and at times nT the impulse acts with propagator K_I given by equation (10.78), so, by the impulse convention and the multiplication rule, the period propagator of the system is

$$K = K_{LO}(T, 0)\, K_1$$

$$= \begin{pmatrix} \cos \omega T & (m\omega)^{-1}\sin \omega T \\ -m\omega \sin \omega T & \cos \omega T \end{pmatrix} \begin{pmatrix} 1 & 0 \\ a & 1 \end{pmatrix}, \qquad (10.124)$$

with trace

$$\tau = 2 \cos \omega T + (a/m\omega) \sin \omega T. \qquad (10.125)$$

For the stability calculation there is no need to obtain the off-diagonal elements of K. The first term comes from the linear oscillator and has modulus 2 when $T = n\pi/\omega$, for positive integer n. Now T is the period of the parametric change and $2\pi/\omega$ is the period of the oscillator, so the oscillator is very sensitive to *any* parametric changes, however small, with periods equal to any multiple of one-half of its natural frequency. Such parametric changes can make the oscillator unstable. For our problem the graph of the trace τ against ωT is shown in figure 10.2.

Fig. 10.2 Stability bands of a linear oscillator with periodic linear impulses. The trace τ of the period propagator K is graphed against ωT, where T is the period of the impulses and $2\pi/\omega$ the oscillator period.

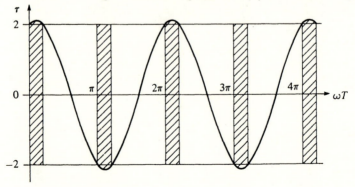

The unstable values of T are given by the values of ωT for which the continuous curve $\tau(\omega T)$ is outside the range $[-2, 2]$. Clearly for the nth instability band one limit is given by $\omega T = n\pi$, and a short trigonometric calculation shows that the other limit is given by

$$\omega T = n\pi + \tan^{-1}(2m\omega/a) \quad (n = 1, 2 \ldots), \qquad (10.126)$$

where the value of the inverse tangent is taken to be between 0 and $\frac{1}{2}\pi$. This instability is known as parametric resonance and occurs, in general, when the period of a periodic parametric change is near any multiple of one-half of the natural period of the unperturbed autonomous system.

Parametric resonance also occurs when the parametric changes are not impulsive, but are smooth periodic functions of time.

Exercises for chapter 10

First-order linear systems

(1) Obtain the motion of a system represented by a function $x(t)$ satisfying the differential equation

$$\dot{x} + cx = a \quad (0 \leqslant t < \infty),$$

where c and a are arbitrary real constants, with initial condition $x(0) = 0$. For what values of the constants is the motion bounded?

(2) The number of births per unit time in a population $x(t)$ of zebras in a game park is $Bx(t)$. The number of zebras that die per unit time is given by $L + Dx(t)$ where L is the number of deaths due to lions and $Dx(t)$ those due to other causes. If x_0 is the population at $t = 0$, determine $x(t)$ for $t > 0$. For what values of the constants B, L and D does the population of zebras increase, decrease or remain steady?

(3) If the constants c, a, T and γ are all positive, obtain and sketch $x(t)$ for systems with equation of motion

$$\dot{x} + cx = b(t), \quad x(-T) = 0,$$

when $b(t)$ is given by

(a) $b(t) = 0$ $(|t| > T)$

       $= a(T - |t|)$ $(|t| \leqslant T)$;

(b) $b(t) = 0$ $(t \leqslant 0)$

       $= a\, e^{-\gamma t}$ $(t > 0)$;

(c) $b(t) = 0$ $\qquad (|t| > T)$

$\qquad = -a(t^2 - T^2) \ (|t| \leqslant T);$

(d) $b(t) = 0$ $\qquad (t \leqslant 0)$

$\qquad = a \sin \gamma t \qquad (t > 0).$

(4) The motion of each of four first-order linear systems is represented by a function $x(t)$ satisfying

$\dot{x} = a(t)x.$

Use propagator theory to determine $x(1)$ when $x(0) = 2,$

(a) if $x(0) = 4$ when $x(1) = 8,$

(b) if $x(\frac{1}{2}) = 1$ when $x(0) = 3,$ and $x(1) = 1$ when $x(\frac{1}{2}) = 3,$

(c) if $x(2) = 6$ when $x(0) = 1,$ and $K(2, 1) = -1,$

(d) if $K(5, -1) = -5,$ $x(1) = 3$ when $x(-1) = 6,$ and $x(0) = 1$ when $x(5) = 10.$

(5) The motion of a system obeys the equation

$\dot{x} = v(x),$

where

$$v(x) = \begin{cases} 0 & |x| > 1 \\ |x| & |x| \leqslant 1 \end{cases}.$$

Find the propagators $K(1, 0), K(0, -1), K(-1, 0), K(1, -1), K(10, 7)$ and $K(10, -10).$ Hence obtain $x(10)$ if $x(-10) = 3.$

(6) The equation of motion of a system is

$$\dot{x} = \begin{cases} 0 & (|t| > T) \\ -a(t^2 - T^2)x & (|t| \leqslant T). \end{cases}$$

Determine the propagators $K(T, -T), K(-T, T), K(2T, -2T)$ and find $x(2T)$ if $x(-2T) = 3.$

(7) The population of finches on an island in the year 1981 is represented by $x(t)$ where t is measured in days from midnight of the last day of 1980. If $B(t)$ is the birth (hatching) rate and $D(t)$ is the death rate per individual bird, then the equation of change for the population is

$\dot{x} = [B(t) - D(t)] \, x.$

It was found that, to a good approximation,

$B(t) = 0$ $\qquad\qquad\qquad (0 \leqslant t < 100, 150 \leqslant t \leqslant 365),$

$B(t) = 2b(t - 100)(150 - t)$ $\quad (100 \leqslant t < 150),$

$D(t) = c + \frac{1}{2} B(t)$ $\qquad\qquad (0 \leqslant t < 365).$

To this approximation, find the propagators $K(100, 0)$, $K(150, 100)$, $K(365, 150)$. If the population was 1000 at the end of 1980 and 1100 at the end of 1981, find a relation between b and c to two significant figures.

(8) Three different systems have an equation of motion of the form $\dot{x} = a(t)x$, where $a(t)$ is of period $T > 0$. Find the period propagator K and determine the stability of motion for all real values of the constants b and c when $a(t)$ is given by

(a) $a(t) = \begin{cases} 0 & (0 \leqslant t < \frac{1}{2}T) \\ b & (\frac{1}{2}T \leqslant t < T), \end{cases}$

(b) $a(t) = b + ct \qquad (0 \leqslant t < T).$

(c) $a(t) = b \,|\, t - \frac{1}{2} T \,| \qquad (0 \leqslant t \leqslant T).$

(9) The equation of motion of a system is $\dot{x} = a(t)x$, where $a(t)$ is defined by

$a(t) = ct \quad (0 \leqslant t < 1),$

$a(t) = a(t + 1).$

Obtain the period propagator K and determine the stability of motion for real non-zero values of c. Obtain also the period propagator $K(t + 1, t)$ for all t and sketch the graph of $x(t)$ between $t = -2$ and $t = 2$ if $x(-2) = 1$. How would the graph change if $x(-2) = 3$?

(10) Prove that if a system has equation of motion $\dot{x} = a(t)x$, where $a(t)$ has period T, then

$K(t + T, t) = K(T, 0),$

so that the period propagator K is independent of the choice of the origin of time, and the stability properties of the system are unaffected by this choice. (Compare this property of first-order systems with the corresponding second-order systems given in exercises 10.19 below.)

(11) The birth and death rates of the birds of exercise 10.7 were observed to depend mainly on the season of the year. Neglecting any other influences, what is the period propagator K of the population, measured at the end of each year? What are $K(365n, 0)$, $K(365n + 100, 100)$, for integer n, assuming years of 365 days, with the day as the unit of time? What are the expected populations at the end of 1990 and the 100th day of 1991?

Exercises on second-order systems

(12) Find the motion of the forced linear oscillator, with Hamiltonian of equation (10.38) in the cases

$$
\text{(a)} \quad F(t) = \begin{cases} 0 & (t \leqslant 0) \\ F_0 & (0 < t < T) \\ 0 & (T \leqslant t), \end{cases}
$$

$$
\text{(b)} \quad F(t) = \begin{cases} 0 & (t \leqslant 0) \\ F_0 t & (0 \leqslant t \leqslant T) \\ 0 & (T < t), \end{cases}
$$

$$
\text{(c)} \quad F(t) = \begin{cases} 0 & (t \leqslant 0) \\ F_0 e^{-\alpha t} & (t > 0), \end{cases}
$$

$$
\text{(d)} \quad F(t) = \begin{cases} 0 & (t \leqslant 0) \\ F_0 e^{-\alpha t} \sin t & (t > 0), \end{cases}
$$

where F_0, α, β and T are positive constants and $q(0) = \dot{q}(0) = 0$.

(13) If in equation (10.49) $\Omega = \omega(1 + \epsilon)$, show that $q(t)$ may be written in the form

$$
q(t) = \frac{2a}{\omega^2(2\epsilon + \epsilon^2)} \ \sin \tfrac{1}{2}\omega\epsilon t \ \sin \omega(1 + \tfrac{1}{2}\epsilon)t.
$$

Sketch the graph of $q(t)$ in the case $\epsilon = 0.1$, showing that $q(t)$ consists of a rapid oscillation of frequency $\omega(1 + \tfrac{1}{2}\epsilon)$, the amplitude of which varies slowly, with frequency $\tfrac{1}{2}\omega\epsilon$.

(14) (a) A linear oscillator is perturbed by the linear time-dependent potential $V(q, t) = -qF(t)$, where $F(t) \to 0$ as $|t| \to \infty$. In the distant past, when $F(t)$ may be neglected, the motion is

$$
q(t) = \left(\frac{2E_0}{m\omega^2} \right)^{\frac{1}{2}} \sin(\omega t + \delta).
$$

Show that, in the future, when $F(t)$ is again negligible, the energy of the oscillator is

$$
E = E_0 + \left(\frac{2E_0}{m} \right)^{\frac{1}{2}} \operatorname{Re}(e^{i\delta}\mathscr{F}) + \frac{1}{2m} |\mathscr{F}|^2,
$$

where

$$
\mathscr{F} = \int_{-\infty}^{\infty} dt\, F(t)\, e^{-i\omega t}.
$$

(b) In the particular case

$$
F(t) = -\epsilon\, U e^{-\lambda |t|} \quad (\lambda > 0),
$$

find $E - E_0$ and compare your result with that of exercise 8.21b. Give a reason for the difference.

(15) Three second-order systems with time-dependent parameters satisfy an equation of motion of the form $\dot{\mathbf{r}} = A(t)\,\mathbf{r}$.

(*a*) For the first system,

when $x(0) = 1, y(0) = 0$, then $x(1) = 2, y(1) = 3$; and

when $x(0) = 0, y(0) = 1$, then $x(1) = -1, y(1) = -1$.

Determine the propagator $K(1, 0)$ from $t = 0$ to $t = 1$ and find $y(1)$ when $x(0) = -1, y(0) = 5$.

(*b*) For the second system, for real, non-zero a, b and T,

when $\mathbf{r}(0) = (a, 0)$, then $\mathbf{r}(T) = (1, -1)$; and

when $\mathbf{r}(0) = (0, b)$, then $\mathbf{r}(T) = (-1, 0)$.

Determine $K(T, 0)$ and find $\mathbf{r}(T)$ when $\mathbf{r}(0) = (1, 1)$.

(*c*) For the third system, for real, non-zero T,

when $\mathbf{r}(0) = (4, 1)$ then $\mathbf{r}(T) = (2, 1)$; and

when $\mathbf{r}(0) = (2, 1)$ then $\mathbf{r}(T) = (0, 1)$.

Determine $K(T, 0)$ and find $\mathbf{r}(0)$ when $\mathbf{r}(T) = (-2, 2)$.

(16) The linear oscillator of example 10.6 is suddenly released at time $t = 0$ to become a free particle. Obtain the propagator $K(t_1, t_0)$ for $t_0 < 0 < t_1$ and find the position $x_1 = x(t_1)$ as a function of $x_0 = x(t_0)$ if $p(t_0) = 0$. For what values of t_0 and x_0 will $p(t_1)$ also be zero for all t_1, and what will the value of x_1 then be?

(17) Find the propagator for a particle of mass m that moves freely, except for two impulses, $P_j(q) = b_j q$ $(j = 1, 2)$, acting at times t'_j $(t'_1 < t'_2)$, where b_j are real non-zero constants.

(18) Find the period propagator K for a particle of mass m that moves freely, except for two sequences of impulses: $P_1(q) = b_1 q$ at times nT; and $P_2(q) = b_2 q$ at times $(n + \frac{1}{2})T$, where b_1 and b_2 are real non-zero constants and n ranges over all the integers. Obtain the eigenvalues of K and discuss the stability properties of the system.

(19) Prove that, if a system has equations of motion $\dot{\mathbf{r}} = A(t)\,\mathbf{r}$, where $A(t)$ has period T, then

$$K(t + T, t) = K(T, 0),$$

so that the period propagator K and the stability properties of the system are independent of the choice of the origin of time. (Compare this result with that of exercise 10.10.)

(20) In the intervals of time

$$nT < t < (n + \rho)T \qquad (T > 0, 0 < \rho < 1),$$

where n ranges over the integers, a particle of unit mass moves freely. In the remaining intervals of time it moves under the action of a linear restraining force, so that its equation of motion is $\ddot{q} = -q$. Determine the values of T for which the motion is stable for $\rho = \frac{1}{2}$ and $\rho = \frac{1}{10}$.

11 CHAOTIC MOTION AND NON-LINEAR MAPS

11.1 Chaotic motion

In the first ten chapters we have presented many analytic theories and methods for the study of dynamical systems and the solution of dynamical problems. The systems were selected so that this analysis could be used.

Most systems are more difficult to understand. For some of them the difficulties can be overcome by a more sophisticated form of the analytic theories that we have described, or by a numerical extension of these analytic methods, but for others they cannot, because the motion is sometimes qualitatively different from anything that we have met before: it is chaotic. The systems do not have to be complicated in order to exhibit chaotic motion. For example, a very natural addition of quadratic or other non-linear terms to one of the linear systems of chapter 10 may represent a system with chaotic motion.

There are analytic theories that throw some light upon chaotic motion, but they are beyond the level of this book, so we have to use numerical methods, supported by some analysis. In practice this is one of the principal ways of studying this phenomenon. Pocket calculators are adequate for the first-order maps of section 11.3; indeed they were used for some of the original discoveries in this field. With the aid of a programmable calculator the reader should have no difficulty in verifying much of this section, and in experimenting on his own, but the second-order area-preserving maps of section 11.4 usually require larger computations.

In chapter 10 we showed the relation between linear area-preserving maps and linear Hamiltonian systems in which the Hamiltonian function is periodic in the time t. In section 11.5 we show that the same applies to non-linear systems, so that chaotic motion for area-preserving maps implies chaotic motion for Hamiltonian systems.

Systems with discrete time whose motion is defined by maps are discussed first of all.

11.2   Maps and discrete time

Time is a continuous variable: Zeno's paradox about Achilles and the Tortoise depended on this continuity; and, for Newton, time was the prime example of a continuous variable.

Nevertheless, we sometimes find it helpful to treat time as if it were discrete. This is particularly true for systems that are affected by conditions that vary periodically with the time. We showed in the preceding chapter that many important properties of linear systems with periodic conditions could be obtained by considering only a sequence of states at times differing by multiples of the period T. This is equivalent to treating the time as a discrete variable. The same applies to non-linear systems with periodic conditions. For example: biological, social and economic systems are affected by the seasons and these vary approximately periodically with a period of one year; for some insects a more appropriate period is the day; parts of a car are subject to periodic forces from the vibration of the engine, and many mechanical, electrical and electronic systems are subject to determinate periodic conditions.

Sometimes the use of a discrete time variable is not a mere matter of convenience, but is obligatory because the data upon which the laws of motion are based may only be available once in a period, because it is too difficult to obtain them at all times, as for some economic and medical data. The approximation of treating differential equations with time as the independent variable as difference equations also requires the time to be treated as discrete.

A system of order n with discrete time is defined by two properties:

(ND1)   The state of the system is represented by n real variables $x_1, x_2, \ldots, x_n$, or one real vector variable $\mathbf{r}$ of dimension n, which may be considered as coordinates of an abstract n-dimensional space named the phase space.

(ND2)   The motion of the system is represented by a sequence of vectors $\mathbf{r}_t$ labelled by the integer time variable t and satisfying the relation

$$\mathbf{r}_{t+1} = \mathbf{F}(\mathbf{r}_t, t), \tag{11.1}$$

where $\mathbf{F}$ is a vector function of $\mathbf{r}$ and t.

We consider only the autonomous systems for which $\mathbf{F}$ does not depend explicitly on the time. The equations of motion are then

$$\mathbf{r}_{t+1} = \mathbf{F}(\mathbf{r}_t) \quad (t = 0, \pm 1, \pm 2, \ldots), \tag{11.2}$$

where $\mathbf{F}$ is a vector-valued function or map of the phase space into itself. In dynamics the word 'map' is usually used in a restricted sense that implies repeated application. For example, it follows from equation (11.2) that

$$\mathbf{r}_1 = \mathbf{F}(\mathbf{r}_0), \quad \mathbf{r}_2 = \mathbf{F}(\mathbf{F}(\mathbf{r}_0)), \quad \mathbf{r}_3 = \mathbf{F}(\mathbf{F}(\mathbf{F}(\mathbf{r}_0))), \text{ etc.} \tag{11.3}$$

For convenience we use the notation $F^{(m)}$ for m repetitions of the map. Thus

$$F^{(1)}(\mathbf{r}) = F(\mathbf{r}), \quad F^{(m)}(\mathbf{r}) = F(F^{(m-1)}(\mathbf{r})) \quad (m = 2, 3, 4, \ldots). \quad (11.4)$$

The index has nothing to do with differentiation. In practice we restrict our attention to non-negative times $t = 0, 1, 2, \ldots$

Frequently systems with discrete time satisfy difference equations, but in that case they can always be brought into the form (11.1) or (11.2). For example, the equation

$$\Delta x_t \equiv x_{t+1} - x_t = G(x_t) \quad (11.5)$$

can be written

$$x_{t+1} = x_t + G(x_t), \quad (11.6)$$

and the equation

$$\delta^2 q_t \equiv q_{t+1} - 2q_t + q_{t-1} = P(q_t) \quad (11.7)$$

can be written

$$q_{t+1} = q_t + p_t + P(q_t)$$
$$p_{t+1} = p_t + P(q_t), \quad (11.8)$$

so it has the form (11.2) with $\mathbf{r} = (q, p)$ and

$$F(q, p) = \begin{pmatrix} q + p + P(q) \\ p + P(q) \end{pmatrix}. \quad (11.9)$$

The equations (11.8) appear in section 11.4, where q represents a configuration coordinate, p a physical momentum and $P(q)$ a periodic impulse.

The reader should compare the theory of this section with the theory of systems of order n with continuous time, presented in section 3.1.

11.3   The logistic map

In section 1.4 we considered the birth and death of the members of a species and this led us to the logistic equation $\dot{x} = bx - cx^2$. This equation can be solved analytically and leads to a stable population $x = b/c$. In order to derive this equation we assumed that the birth and death rates did not depend explicitly upon the time. This is a reasonable assumption for bacteria and humans, but is very bad for birds and the many animals and insects that have definite breeding seasons. It then makes sense to label each breeding season by the discrete variable $t = 0, 1, 2, \ldots$ (which usually has the year as its unit), with an arbitrarily chosen zero, and consider the population, N_t, just after the breeding is complete. The population is supposed to be sufficiently large for N to be a continuous variable.

As in section 1.4 for the logistic process, the change in population has a linear term primarily due to birth and a quadratic term due to death, since the death rate per individual tends to increase as the population becomes sufficiently large.

The population N_t then satisfies the *logistic mapping*† equation

$$N_{t+1} = N_t(b - cN_t) \quad (b > 1, c \geqslant 0). \tag{11.10}$$

For $c = 0$, or $c \neq 0$ with the limit of small N, it represents a population that grows exponentially as a result of reproduction. Otherwise, for $c \neq 0$, the population is limited, as it is in practice by shortages of food, disease due to overcrowding and other factors. For the logistic form (11.10) the reduction due to the quadratic term becomes drastic as N_t approaches b/c and the behaviour is then very different from the solution of the logistic equation with continuous time.

To study this system, we first normalize N to give the normalized population

$$x = c N/b, \tag{11.11}$$

bringing the equation to the form

$$x_{t+1} = b x_t(1 - x_t). \tag{11.12}$$

This is of the standard form (11.2) of a first-order system with discrete time with

$$F(x) = F_b(x) = bx(1 - x). \tag{11.13}$$

It has been shown that this very simply defined non-linear system has the same universal properties as many other first-order systems possessing a smooth mapping function $F(x)$ with just one maximum and no other stationary points.

The logistic map (11.12) can easily be studied by numerical experiment using a pocket calculator. For x_t in the interesting range $0 \leqslant x_t \leqslant 1$, it exhibits an extraordinarily subtle and varied behaviour for different values of the constant b lying in the range $1 < b \leqslant 4$.

Some of this behaviour is exhibited in figure 11.1 for $b = 2, 3.1, 3.5, 3.6$ and 3.84, with initial condition $x_0 = 0.2$ in every case.

For $b = 2$ the normalized population, x_t, tends rapidly towards its equilibrium value at $x = 0.5$, just as for the logistic differential equation (1.21). This strongly stable fixed point is the only attractor, and it attracts all motion in the interval $0 < x < 1$.

For $b = 3.1$ there is no single fixed point. After a few generations, or values of t, the population oscillates between the neighbourhoods of the two values $x = 0.557$ and $x = 0.765$. These two points together form an attractor, and the motion between them is periodic, of period 2. This motion is stable.

---

† see May, R.M. (1976). Simple mathematical models with very complicated dynamics. *Nature*, **261**, 459–467.

Fig. 11.1 Motion of a system satisfying the logistic mapping equation (11.12) for $b = 2, 3.1, 3.5, 3.6$ and 3.84, with initial conditions $x_0 = 0.2$ in each case.

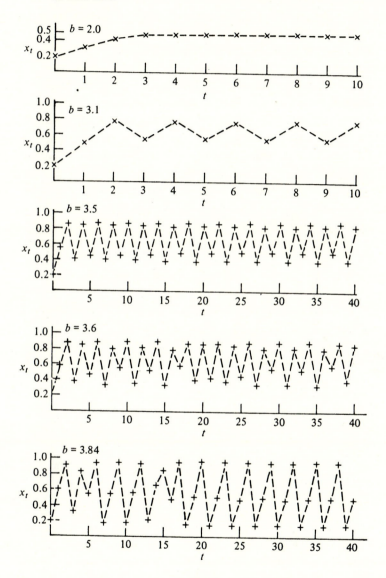

For $b = 3.5$ the behaviour is different from that for either of the previous values of b. After a sufficient number of generations, more than before, the motion approaches close to a periodic motion of period 4. This periodic motion is stable and the attractor consists of four points. By slightly increasing the value of b, values of b can be found for which the motion is attracted to stable cycles of period 2^n for an arbitrary positive integer n.

However, before b reaches 3.6, no stable periodic motion remains. For $b = 3.6$ our numerical experiment shows no periodic or even regular behaviour even for 40 generations, and it has been established that there is a significant set of values of b for which the motion *never* settles down to a periodic or even a regular pattern of behaviour. Such motion is known as *chaotic* or *irregular* and is a common, though not universal, feature of complex dynamical systems.

For the logistic map, the chaotic motion defies simple analysis, despite the simplicity of the equations defining the motion. There is no escape from the conclusion that populations with a definite reproductive season, with no more than one generation per season, show very complicated behaviour over long periods of time, even though the equations defining their change have a very simple form.

This is true of many processes, biological, physical or economic, that are governed by a mapping function like (11.13) with a maximum for some values of multiplying parameter b. Chaotic change of populations of birds and insects is commonly observed.

The subtlety of the population model (11.12) is not yet exhausted. When b is increased to 3.84 the population settles down fairly quickly to a motion of period 3 as shown in figure 11.1. Increasing b still further within the permitted range $1 < b \leqslant 4$ can produce chaos again, and so on.

How are we to understand all of this? It is not yet fully understood, and is still an active area of research, but some properties can be studied by elementary methods.

The fixed points of the map generated by the function $F_b(x)$ are the solutions of

$$x = F_b(x) \tag{11.14}$$

given by the intercepts of the graphs of the left and right sides of the equation. This is illustrated for the quadratic map (11.12) at the top of figure 11.2 (p.204) for two values of b that were the subject of our numerical experiments. In this case equation (11.14) is a quadratic equation with solutions $x^{(0)} = 0$ and $x^{(1)} = 1 - 1/b$. The solution $x^{(0)}$ is trivial. The numerical experiments suggest that the solution $x^{(1)}$ is a stable fixed point for $b = 2$ and an unstable fixed point for $b = 3.1$. The range of b where $x^{(1)}$ is stable is given by a Taylor expansion of $F_b(x)$ in its neighbourhood:

$$F_b(x) = x^{(1)} + F_b'(x^{(1)})\,(x - x^{(1)}) + O(x - x^{(1)})^2.$$ (11.15)

For points sufficiently close to $x^{(1)}$, the quadratic term may be neglected and the map $F_b(x)$ is approximated by a linear map whose stability is given in section 10.2 by the inequality (10.31). The point

$$x^{(1)} \text{ is } \left\{ \begin{array}{c} \text{stable} \\ \text{unstable} \end{array} \right\} \text{ if } \left\{ \begin{array}{c} |F_b'(x^{(1)})| < 1 \\ |F_b'(x^{(1)})| > 1 \end{array} \right\}.$$ (11.16)

In general the slope of the graph of the mapping function at any fixed point determines the stability at that fixed point. If the slope is less than unity in magnitude, the fixed point is stable, and if it is greater than unity, the fixed point is unstable. This is illustrated by the tangents to $F_b(x)$ at $x^{(1)}$ for $b = 2$ and $b = 3.1$ in figure 11.2, showing stability and instability respectively. For the logistic map we have $x^{(1)} = 1 - b^{-1}$ and

$$|F_b'(x^{(1)})| = |b(1 - x^{(1)})| = |b - 2|,$$ (11.17)

so the fixed point $x^{(1)}$ is stable for $b < 3$. It is unstable for $b > 3$, so what happens then?

It is helpful to look at the *second generation map* defined by the function

$$F_b^{(2)}(x) = F_b(F_b(x)),$$ (11.18)

representing the population after two generations. The functions $F_2(x)$, $F_{3.1}(x)$ appear in figure 11.2*a* and $F_2^{(2)}$ and $F_{3.1}^{(2)}$ in figures 11.2*b* and 11.2*c*.

For any value of b, the fixed points of F_b are also fixed points of $F_b^{(2)}$, but the converse is not true and $F_b^{(2)}$ sometimes has fixed points that are not fixed points of F_b. They are the points of cycles of period 2. The derivative of $F_b^{(2)}$ at $x^{(1)}$ is found, by the chain rule, to be

$$\frac{dF_b^{(2)}}{dx}(x)\Big|_{x = x^{(1)}} = \frac{d}{dx}F_b(F_b(x))\Big|_{x = x^{(1)}}$$

$$= F_b'(x^{(1)})\,F_b'(F(x^{(1)}))$$

$$= F_b'(x^{(1)})^2.$$ (11.19)

The value of b for which F becomes unstable at $x = x^{(1)}$ is a value for which $F^{(2)}$ has unit slope at $x^{(1)}$. But it is clear from figures 11.2*b* and 11.2*c* that, if the slope of $F^{(2)}$ at $x^{(1)}$ is less than unity, then the equation $x = F^{(2)}(x)$ has one solution and $F^{(2)}$ has one fixed point at $x^{(1)}$ whereas, if the slope is greater than unity, $F^{(2)}$ has two additional fixed points at $x^{(1)}$.

The two new solutions $x^{(2)}$ and $x^{(3)}$ are fixed points of $F^{(2)}$, but not of F, so they must belong to a cycle of period 2. Furthermore, figure 11.2*c* shows that the slopes of $F^{(2)}$ at $x^{(2)}$ and $x^{(3)}$ are less than unity, so the 2-cycle is stable. So, at the value of b where the fixed point $x^{(1)}$ becomes unstable, a stable 2-cycle is produced. Such a change in the nature of the stable motion is known as a *bifurcation*, implying a structure like that of a fork, as illustrated later in figure 11.3.

Fig. 11.2 (*a*) Graphs of x and $F_b(x)$ for the logistic mapping equation (11.12) with b = 2, 3.1, illustrating stability and instability of fixed points.
(*b*) and (*c*) Graphs of the second generation mapping function $F_b^{(2)}(x)$ for b = 2 and 3.1 respectively.

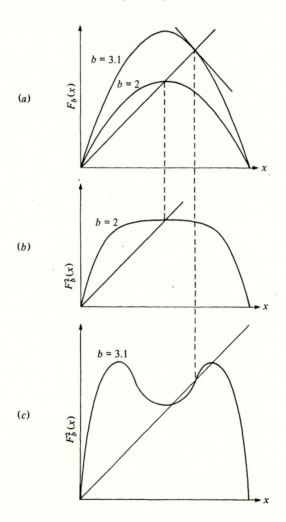

Fig. 11.3 Stable and unstable fixed points as functions of b, illustrating bifurcation structure.

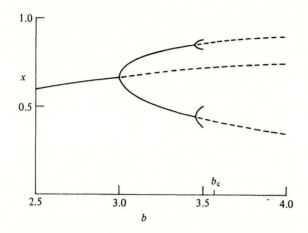

As the value of b increases still further, the slopes of $F^{(2)}$ at $x^{(2)}$ and $x^{(3)}$ decrease until they are equal to -1. A further bifurcation then occurs, the 2-cycle becomes unstable and a stable 4-cycle appears. This is repeated for 2^n-cycles with all positive integers n, with the change in b after each bifurcation decreasing so rapidly with n that beyond a critical value b_c of b all the 2^n-cycles are unstable. It is found that

$$b_c = 3.5700 \ldots \tag{11.20}$$

For $b > b_c$ the chaotic behaviour commences. In figure 11.3 the fixed points of the 1-cycle $x^{(1)}$ and of the 2-cycle $x^{(2)}$ and $x^{(3)}$ are illustrated as functions of b, clearly showing the bifurcations, together with a small section of the 4-cycle.

Further numerical experiment is needed to investigate higher values of b, where 3-cycles and stable and unstable periodic motion of all other periodicities is found before b reaches 4.

11.4   Quadratic area-preserving maps

There are Newtonian or Hamiltonian systems with chaotic motion. The simplest have a periodic sequence of impulses and are discussed in this section. Other Hamiltonian systems are briefly described in the next section.

In order to study fundamental particles experimentally, physicists need to keep them rotating around storage rings at a nearly constant rate for long periods of time, as shown in the diagram. A particle may make more than 10^{10} revolutions around the ring, so the displacement of the particle must be kept small during all these revolutions or it will collide with the walls and be lost.

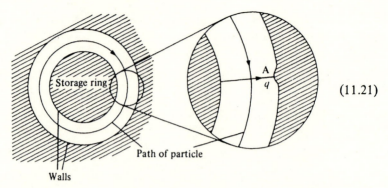

$$(11.21)$$

Any local imperfection in the wall, such as that shown at the point A, can produce an inhomogeneous magnetic field that gives an impulse to the particle, deflecting it from its best path. The impulse is a highly non-linear function of the displacement q, measured when the particle passes A. Since the impulse may be applied 10^{10} times, a very small imperfection can result in displacements sufficient to deflect almost all the particles to the walls, making the storage ring useless. Whether or not this happens depends on the stability of the motion. Because of the possibility of chaotic motion, this is much more subtle than the stability of linear systems considered in chapter 10, but the theory of that chapter is very helpful.

Similar problems arise in the theory of the motion of the planets of the Solar System, as they have rotated around the Sun of the order of 10^9 times, fortunately for us without the inner planets, such as the Earth, being ejected from the system by the major planets, such as Jupiter.

The theory also applies to the containment of charged particles by magnetic fields in devices constructed to produce energy by nuclear fusion; to the motion of charged particles in the Earth's magnetic field, as in the radiation belts that produce the Aurorae; to the motion of atoms in polyatomic molecules; and to many other situations. The general theory is too complicated for us, so we restrict our attention to a simple example that shows many of the important phenomena and closely models the storage ring stability problem.

A particle has unit mass and one degree of freedom, with coordinate q and conjugate momentum p. It moves freely almost all of the time, but at every integer value of the time, t, it is subject to an impulse $P(q)$. That is

$$\text{impulse} = P(q) \quad (t = 0, 1, 2, 3, \ldots) \qquad (11.22)$$

where the value of P depends upon q in a known way, but not upon the value of the integer t. We use the notation

$$q_t = q(t), p_t = p(t) \quad (t = 0, 1, 2, 3, \ldots) \qquad (11.23)$$

for the coordinate and momentum at non-negative integer times t. By the impulse convention (10.81) the values $p(t)$ are those just before the impulse is applied.

The equations of motion can easily be solved for the unit time intervals of free motion to give the second-order map from the state at time t to the state at time $t + 1$ given by

$$q_{t+1} = q_t + p_{t+1} = q_t + p_t + P(q_t) \tag{11.24a}$$

$$p_{t+1} = p_t + P(q_t). \tag{11.24b}$$

Since the free motion and the impulse both produce area-preserving, or canonical, transformations, the combined transformation (11.24) represents an area-preserving map from the phase space (q, p) onto itself. There are fixed points where

$$p = 0, \quad P(q) = 0 \tag{11.25}$$

and we choose the origin at one of these fixed points. The coordinate as a function of time then changes as shown in diagram (11.26). The best orbit of

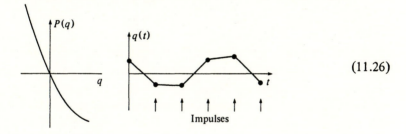

$$(11.26)$$

the storage ring problem passes the point A at a zero of $P(q)$. Now we expand $P(q)$ about this fixed point as the origin.

The linear term $P(q) = aq$ is insufficient to show the principal features of the motion but it can be used to determine the stability of the fixed point at the origin, as in example 10.8. It is stable when $-4 < a < 0$, so we choose a in this range in order to study non-linear motion around a stable fixed point.

The principal features of the motion do appear when the quadratic term is included, so we truncate the expansion there and consider the quadratic impulse

$$P(q) = aq + bq^2 \quad (-4 < a < 0). \tag{11.27}$$

In example 10.8 we found it convenient to transform to coordinate (X, Y), such that the propagator became a rotation matrix $L(-\alpha)$ with

$$\cos \alpha = 1 + \tfrac{1}{2}a \quad (\alpha > 0, -4 < a < 0), \tag{11.28}$$

as in equation (10.121). The same transformation can be applied to the system with the quadratic impulse (11.27) and the linear term is then reduced to standard form, but the quadratic term becomes more complicated.

However, a further change of scale, or dilation, and a reflexion in a straight line through the origin, result in the standard quadratic area-preserving Cremona map of Siegel and Hénon[†]. In the new representation (x, y) it is given by

$$x_{t+1} = x_t \cos \alpha - (y_t - x_t^2) \sin \alpha$$
$$y_{t+1} = x_t \sin \alpha + (y_t - x_t^2) \cos \alpha. \qquad (11.29)$$

This can be written

$$\mathbf{r}_{t+1} = \mathbf{F}_\alpha(\mathbf{r}_t) \quad (\mathbf{r} = (x, y)), \qquad (11.30)$$

where α is the angle of rotation corresponding to the linear part of the map; this is now positive because the reflexion reverses the sense of rotation. Almost any quadratic area-preserving map can be transformed by a non-singular linear transformation to the form (11.29), so the structure of the motion depends on one positive parameter only, the rotation angle α.

We know from section 10.6, starting at equation (10.105), that the linear part of the map (11.29) is structurally unstable for arbitrarily small changes in α. This is also true of the quadratic map, but the effects are much more drastic.

As in the linear case the motion can be represented by the series of points $\mathbf{r}_t$ in the $\mathbf{r} = (x, y)$ plane. Using digital computers, large numbers of points can be produced very quickly and plotted automatically. Such plots enable us to get a good qualitative picture of what is happening over large areas of the phase plane for different values of α. In figures 11.4–11.7 we reproduce four such plots of Hénon, each with points belonging to several different initial conditions for a given value of $\cos \alpha$.

The motion is extraordinarily complicated. For a given initial point $\mathbf{r}_0$ close to the origin, the motion is usually confined to an invariant curve. Those closest to the origin resemble the invariant curves of the linear motion in that neighbourhood, but further from the origin they become more distorted.

The motion on the invariant curve consists of a sequence of points regularly spaced around it, and is known as *regular* motion. The curves resemble the invariant curves of a Hamiltonian system with continuous time t which are the contours of the Hamiltonian as described in chapter 4.

---

† Hénon, M. (1969). Numerical study of quadratic area-preserving mappings. *Quarterly of Applied Mathematics,* **27**, 291-312.

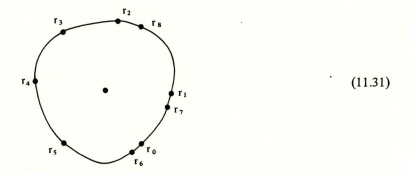

(11.31)

Furthermore, it is possible to find an angle variable for an invariant curve so that the motion on it has the form

$$\mathbf{r}_t = \mathbf{f}(\omega t + \delta) = \mathbf{f}(\theta) \quad (t = 0, 1, 2, \ldots), \tag{11.32}$$

where $\mathbf{f}$ is some smooth vector function of an angle variable θ, as in chapter 7. The only difference is that the time takes integer values only, so that the motion consists of a sequence of discrete points around the curve instead of a continuous motion around it. The angular frequency ω decreases from a value near α for small curves near the origin to smaller values for the large distorted curves, just as in the case of continuous time.

Fig. 11.4 Phase points generated by the map (11.29) for $\cos \alpha = 0.8$.

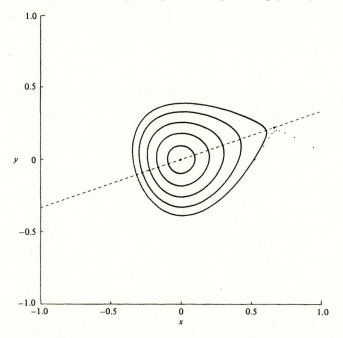

For cos α = 0.8 in figure 11.4, we see some of these invariant curves. When
the initial point is outside some critical invariant curve, the points r_t are seen to
scatter and then to diverge to infinity. Similar, but continuous, divergence is
seen for the linear repulsive force of example 4.3, but this similarity is
deceptive. For discrete time there are completely new phenomena for almost all
values of α. These are not observed for cos α = 0.8 because they are not very
prominent.

However, for cos α = 0.24 in figure 11.5 we observe a difference. Because
$α/2π = 0.211$ is slightly greater than $\frac{1}{5}$, there is a stable cycle of five elliptic
points, each a fixed point of $F^{(5)}$, though not of $F^{(1)}$. Around these five points
appears a chain or necklace of five *islands*, each containing a set of invariant
curves surrounding one of the fixed points. These five islands appear in the

Fig. 11.5 Phase points generated by the map (11.29) for cos α = 0.24.

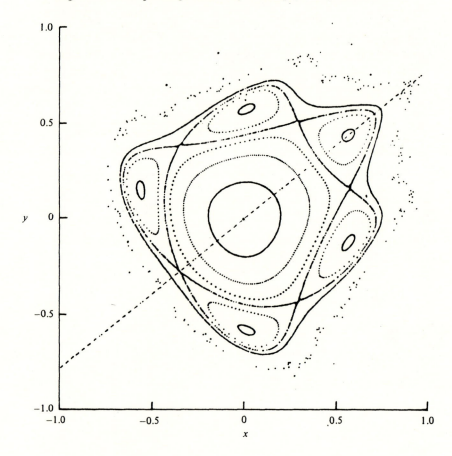

region of phase space between the invariant curves with irrational winding numbers, $\omega/2\pi$, slightly greater, and slightly less, than $\frac{1}{5}$. More detailed calculations show similar chains of islands around stable points of $F^{(D)}$ for arbitrary positive integer D, between the invariant curves with winding numbers in the neighbourhood of any rational N/D. Since there is an infinite number of rationals between any two irrationals, there is an infinite number of those chains of islands, but those corresponding to long periods D are very small and difficult to observe unless specifically looked for in the numerical calculations.

For $\cos \alpha = 0.24$ there is also an unstable 5-cycle of hyperbolic points at a similar distance from the origin. They appear to be joined by a separatrix, but this is deceptive. Figure 11.6 shows a single orbit of successive phase points near the same unstable 5-cycle for $\cos \alpha = 0.22$. The orbit appears to reach arbitrarily close to all points of a region of finite area in the phase space. The successive points in this orbit behave in a chaotic fashion, similar to the chaotic motion of the first-order system of the previous section, by contrast to the regular motion

Fig. 11.6 Phase points generated by the map (11.29) for $\cos \alpha = 0.22$.

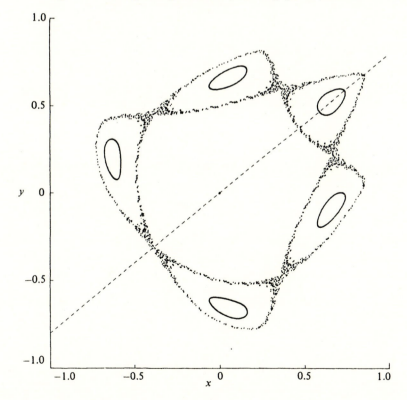

on the invariant curves. It is typical of motion in the neighbourhood of any
unstable periodic orbit, with rare exceptions. Chaotic motion is present for
$\cos \alpha = 0.8$ and 0.24, but is difficult to see in figures 11.4 and 11.5 because it
occupies very narrow regions. In some systems the motion in a large part of
phase space is chaotic.

There are thus three distinct, significant types of motion for the quadratic
area-preserving map. There is regular motion on invariant curves around the
fixed points of stable periodic orbits, there is chaotic motion bounded within
regions of finite area around unstable periodic orbits and there is unbounded
escaping motion. The invariant regions of regular and chaotic motion are mixed
together in a very complicated way. If we look closely at a small region of phase
space as in figure 11.7, we see that the pattern of chains of islands is repeated on
smaller and smaller scales. These smaller chains surround periodic orbits of
longer periods. Periodic motions of different periods coexist and their stability
properties guide us through a sea of confusing change.

Fig. 11.7 An enlargement of a small part of figure 11.5.

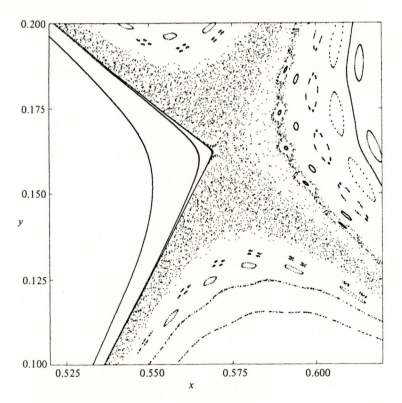

The mathematical analysis of chaotic motion has not reached the stage where it can handle all the properties that have been observed in numerical experiments. The field is still developing.

So far we have not considered second-order maps that do not preserve area. These can have attractors, like the one-dimensional maps and the second-order systems with continuous time. Like the latter the attractors may be stable points or limit cycles as in chapter 3, or they may be a new type, named *strange attractors*, that support chaotic motion.

11.5   Regular and chaotic motion of Hamiltonian systems

In chapter 10 we saw how the theory of systems satisfying linear differential equations with periodic velocity functions led naturally to the study of maps. Now we show that this is true for non-linear systems.

Consider a system of order n whose motion satisfies a differential equation of the standard form

$$\frac{d\mathbf{r}}{dt} = v(\mathbf{r}, t),$$
(11.33)

where $v(\mathbf{r}, t)$ is periodic in time, with the same period for any fixed $\mathbf{r}$, and bounded throughout $(\mathbf{r}, t)$-space. Choose the unit of time to be the period, giving

$$v(\mathbf{r}, t + 1) = v(\mathbf{r}, t).$$
(11.34)

As stated in appendix 1, the initial condition $\mathbf{r}(0)$ determines the motion $\mathbf{r}(t)$ for all time and, in particular, it determines $\mathbf{r}(1)$. The differential equation (11.33) thereby defines a map $\mathbf{F}$ from the phase space onto itself so that

$$\mathbf{r}(1) = \mathbf{F}(\mathbf{r}(0))$$
(11.35)

for every solution $\mathbf{r}(t)$ of equation (11.33).

But, for every solution $\mathbf{r}(t)$,

$$\frac{d}{dt}\mathbf{r}(t + 1) = \frac{d}{d(t + 1)}\mathbf{r}(t + 1)$$

$$= v(\mathbf{r}(t + 1), t + 1)$$

$$= v(\mathbf{r}(t + 1), t),$$
(11.36)

by the periodicity condition (11.34), so $\mathbf{r}(t + 1)$ is also a solution, and, by induction, so is $\mathbf{r}(t + s)$ for every positive integer s. We substitute into equation (11.35) and find that

$$\mathbf{r}(s + 1) = \mathbf{F}(\mathbf{r}(s)) \quad (s = 0, 1, 2, \ldots).$$

On replacing s by integer t and using the notation $\mathbf{r}_t = \mathbf{r}(t)$, we obtain the standard form of the equation of motion for an autonomous system with discrete time

$$\mathbf{r}_{t+1} = \mathbf{F}(\mathbf{r}_t). \tag{11.38}$$

So we find that a non-autonomous system with continuous time and differential equations of motion with periodic velocity function leads to the mapping equation typical of an autonomous system with discrete time.

The solution of the differential equations of motion can be divided into two stages:

Stage 1: The derivative of the map $\mathbf{F}$ by the general solution of the differential equation of motion from $t = 0$ to $t = 1$.

Stage 2: The solution of the mapping equation generated by the map $\mathbf{F}$.

For systems of order 2 or more, the mapping equation may lead to chaotic motion, which will then be present in the solutions of the original differential equations. For example, systems of order 2 or more whose motion satisfies a differential equation of standard form (11.33) with periodic conditions can exhibit chaotic motion. Furthermore this motion is often important. This is the case for Hamiltonian systems of one degree of freedom, with a Hamiltonian function $H(q, p, t)$ depending periodically on t.

Where there is bounded motion, a significant proportion of phase space is usually occupied by phase points which undergo regular motion and a significant proportion by phase points for which the subsequent motion is chaotic. The same applies to *conservative* Hamiltonian systems of more than one degree of freedom. This limits the practical applicability of the methods described in this book.

However, there are many such systems for which the regular motion is important and some, the so-called *integrable systems*, for which it is the only motion present. For these systems, the methods of the first ten chapters can be used, but the additional degrees of freedom sometimes lead to additional difficulties, particularly for perturbation theory and the theory of adiabatic invariance; these difficulties are beyond the scope of this book.

Exercises for chapter 11

These exercises are best solved using a computational aid such as a pocket calculator, preferably programmable.

(1) Obtain and graph the motion x_t for the logistic map (11.12), with $b = 1.5$ from $x_0 = 0.1$ and from $x_0 = 0.8$ to the time when x_t reaches within 0.001 of $x^{(1)}$.

(2) Find the points $x^{(3)}$ and $x^{(4)}$ of the stable 2-cycle of period 2 of the logistic map (11.12) with $b = \frac{10}{3}$, to three decimal places, by obtaining the motion from the initial point $x_0 = 0.95$, and verify that the same

cycle is approached when $x_0 = 0.35$. What happens when $x_0 = 0.7$ and $x_0 = 0.701$, and why? The last motion need be evaluated to $t = 6$ only.

(3) Eliminate the factor $(x - x^{(0)})$ from the quadratic equation $x - F_b^{(2)}(x) = 0$ for the logistic map (11.12) and eliminate the root $x^{(1)}$ from the resultant cubic equation to obtain a quadratic equation. Find the roots of this equation and hence obtain the stable periodic orbit for $b = \frac{10}{3}$ to three decimal places. Check with the answer to the previous exercise, if it has been obtained.

(4) The map $F(x)$ has the second generation map $F^{(2)}(x)$ and a cycle of period 2 with fixed points y and z. Prove that

$$F^{(2)\prime}(y) = F^{(2)\prime}(z) = F\,'(y)F\,'(z),$$

where the prime denotes differentiation. Hence find the minimum value b_2 of b, such that the 2-cycles of the logistic map (11.12) become unstable when $b > b_2$. You may assume that y and z are given by $(b + 1 \pm \sqrt{(b-3)(b+1)})/2b$.

(5) Obtain a condition for stability of an m-cycle $y_1, y_2, \ldots, y_m$, of the map $F(x)$ in terms of the derivatives $F'(y_k)$.

(6) Two successive generations of a biological population are related by the equation

$$x_{t+1} = x_t \exp[r(1 - x_t)] \quad (r > 0).$$

Like the logistic map (11.12) this describes a population with exponential growth at low densities, and represents the effect of epidemic disease at high densities particularly well. It is more complicated than the quadratic map, but has the advantage that a positive population stays positive. Find the value of r at which the fixed point becomes unstable, and find the stable motion and its period for $r = 2.3$ and 2.6, by numerical experiment.

Note that these numerical experiments enable us to find solutions of transcendental equations by an iterative procedure, provided certain stability criteria are satisfied.

APPENDIX 1
EXISTENCE THEOREMS

There are many forms of existence theorem for differential equations, whose proofs alone could occupy a book of this size. We state three theorems: A1, A2 and A3. Theorems A1 and A2 are most useful; theorems A2 and A3 have been proved by Arnold (1973).[†] At the end of the appendix we show how theorem A1 follows from theorem A3.

In all these theorems the vector fields v are differentiable and the derivatives are continuous.

Theorem A1

Suppose that $\mathbf{r}$ is a point in an n-dimensional Euclidean vector space $\mathbf{R}^n$ and $v(\mathbf{r}, t)$ is a vector field that satisfies

$$|\, v(\mathbf{r}, t)\,| < v_m, \tag{A1.1}$$

for all real t in the interval

$$I_t: |\, t - t_0\,| < T \tag{A1.2}$$

and all points $\mathbf{r}$ of the ball

$$S_r: |\, \mathbf{r} - \mathbf{r}_0| < v_m\, T = R, \tag{A1.3}$$

for some fixed

$$\mathbf{r}_0 \in \mathbf{R}^n, t_0 \in \mathbf{R} \quad (T > 0, v_m > 0). \tag{A1.4}$$

Then the differential equation

$$\frac{d\mathbf{r}}{dt} = v(\mathbf{r}, t) \quad (\mathbf{r} \in \mathbf{R}^n, t \in \mathbf{R}), \tag{A1.5}$$

with condition

$$\mathbf{r}(t_0) = \mathbf{r}_0 \tag{A1.6}$$

has a unique solution $\mathbf{r}(t)$ in the interval I_t.

[†] Arnold, V.I. (1973). *Ordinary Differential Equations*. Massachusetts: M.I.T. Press.

Theorem A2

The solution of the equation

$$\frac{dx}{dt} = v(x, t) \quad (x \in M, t \in R),\tag{A1.7}$$

defined by the vector field v on a *compact* manifold M, with initial condition $x(t_0) = x_0$, has a unique solution for *all* real time t.

A circle, sphere and torus are all compact manifolds, so a system whose phase space is a torus, for example, and whose phase velocity function represents a vector field, has an equation of motion of the form (A1.7) whose solution exists for all time. There is no terminating motion.

Proofs

We do not provide proofs, but, for completeness, we indicate how they can be found. Theorem A2 is proved by Arnold (1973) in his section 35.1. This section also contains the proof of a theorem of which the following is a special case.

Theorem A3

If $\rho \in R^m$ and $u(\rho)$ is a vector field that is non-zero only in a compact subset K of R^m, then the differential equation

$$\frac{d\rho}{dt} = u(\rho),\tag{A1.8}$$

with condition

$$\rho(t_0) = \rho_0 \quad (\rho_0 \in R^m, t_0 \in R)\tag{A1.9}$$

has a unique solution for all real t.

To prove theorem A1, choose any $\epsilon > 0$ and apply theorem A3 to the vector field

$$u(r, s) = (v(r, s) F_\epsilon(|r - r_0|/R), F_\epsilon(|t - t_0|/T), (r, s) \in R^{n+1}\tag{A1.10}$$

where $F_\epsilon(x)$ is a smooth function satisfying

$$\begin{aligned}F_\epsilon(x) &= 1 \quad (x \leqslant 1 - \epsilon),\\ F_\epsilon(x) &= 0 \quad (x \geqslant 1).\end{aligned}\tag{A1.11}$$

Such a function is defined by (A1.11) with

$$F_\epsilon(x) = \frac{1}{2}\left[1 - \tanh\left(\frac{1}{1-x} + \frac{1}{1-\epsilon-x}\right)\right] \quad (1 - \epsilon < x < 1).\tag{A1.12}$$

The subset K is

$$K: |r - r_0| < R, \quad |t - t_0| < T\tag{A1.13}$$

and the differential equations to be satisfied are

$$\frac{\mathrm{d}s}{\mathrm{d}t} = F_\epsilon(|t - t_0|/T),$$

$$\frac{\mathrm{d}\mathbf{r}}{\mathrm{d}t} = v(\mathbf{r}, s) F_\epsilon(|\mathbf{r} - \mathbf{r}_0|/R).$$

$$(A1.14)$$

We choose the initial condition ρ_0 so that

$$\mathbf{r}(t_0) = \mathbf{r}_0, \ s(t_0) = t_0. \tag{A1.15}$$

By theorem A3, there is a unique solution for all t. From the form of the equations, this is also the unique solution of (A1.5), with condition (A1.6), provided t is in the interval $I_{t\epsilon}$ and $\mathbf{r}$ is in the ball $S_{r\epsilon}$ defined by

$$I_{t\epsilon}: |t - t_0| < T(1 - \epsilon),$$

$$S_{r\epsilon}: |\mathbf{r} - \mathbf{r}_0| < R(1 - \epsilon). \tag{A1.16}$$

But, from the inequality

$$|\mathbf{r}(t) - \mathbf{r}_0| = \left| \int_{t_0}^{t} \mathrm{d}t \, v(\mathbf{r}(t), t) \right| \leqslant \left| \int_{t_0}^{t} \mathrm{d}t \, |v| \right|$$

$$< v_m \left| \int_{t_0}^{t} \mathrm{d}t \right| = v_m |t - t_0|, \tag{A1.17}$$

$\mathbf{r}(t)$ is always in $S_{r\epsilon}$ when t is in $I_{t\epsilon}$ so there is a unique solution of (A1.5) with condition (A1.6) for t in $I_{t\epsilon}$, and, since ϵ may be arbitrarily small, in I_t.

APPENDIX 2

INTEGRALS REQUIRED FOR SOME SOLUBLE PROBLEMS

A number of potentials produce motions which can be described in terms of elementary functions. Nevertheless, the evaluation of the necessary integrals can be very tedious. Here we give the relevant integrals for the four standard potentials:

$$V_1(q) = U \tan^2 \alpha q \tag{A2.1}$$

$$V_2(q) = U(e^{-2\alpha q} - 2e^{-\alpha q}) \tag{A2.2}$$

$$V_3(q) = -U \operatorname{sech}^2 \alpha q \tag{A2.3}$$

$$V_4(q) = U\left[\left(\frac{\alpha}{q}\right)^2 - 2\left(\frac{\alpha}{q}\right)\right], \tag{A2.4}$$

where α and U are positive constants.

The definite integrals needed to obtain the actions of the motion in these potentials are:

$$J_1 = \int_0^{q_1} dq\, (E - U \tan^2 \alpha q)^{\frac{1}{2}} \quad (\tan \alpha q_1 = (E/U)^{\frac{1}{2}})$$

$$= \pi(\sqrt{E+U} - \sqrt{U})/2\alpha \quad (E > 0). \tag{A2.5}$$

$$J_2 = \int_{q_1}^{q_2} dq\, [E - U(e^{-2\alpha q} - 2e^{-\alpha q})]^{\frac{1}{2}}$$

$$= \pi(\sqrt{U} - \sqrt{-E})/\alpha \quad (-U < E < 0), \tag{A2.6}$$

where

$$q_{1,2} = \frac{1}{\alpha} \ln \left[\frac{\sqrt{U}}{\sqrt{U} \mp \sqrt{E+U}}\right].$$

$$J_3 = \int_0^{q_1} dq \, (U \operatorname{sech}^2 \alpha q + E)^{\frac{1}{2}} \qquad (\cosh \alpha q_1 = (-U/E)^{\frac{1}{2}})$$

$$= \pi(\sqrt{U} - \sqrt{-E})/2\alpha \qquad (-U < E < 0). \tag{A2.7}$$

$$J_4 = \int_{q_1}^{q_2} dq \left[E - U \left(\left(\frac{\alpha}{q} \right)^2 - 2 \left(\frac{\alpha}{q} \right) \right) \right]^{\frac{1}{2}}$$

$$= \pi \alpha \sqrt{U} \, [\sqrt{-U/E} - 1] \qquad (-U < E < 0), \tag{A2.8}$$

where

$$q_{1,2} = \frac{\alpha}{E} \, (\pm \sqrt{U^2 + EU} - U).$$

To obtain the time as a function of the coordinate q, the following indefinite integrals are required.

$$K_1(q) = \int dq \, (E - U \tan^2 \alpha q)^{-\frac{1}{2}}$$

$$= \frac{1}{\alpha \sqrt{E + U}} \sin^{-1} \left[\sqrt{\frac{U + E}{E}} \sin \alpha q \right]. \tag{A2.9}$$

$$K_2(q) = \int dq \, [E - U (e^{-2\alpha q} - 2 e^{-\alpha q})]^{-\frac{1}{2}}$$

$$= \frac{1}{\alpha \sqrt{-E}} \sin^{-1} \left[\frac{-E e^{\alpha q} \, U}{\sqrt{U(U + E)}} \right] \qquad (-U < E < 0), \tag{A2.10a}$$

$$= \frac{1}{\alpha \sqrt{E}} \cosh^{-1} \left[\frac{E e^{\alpha q} + U}{\sqrt{U(U + E)}} \right] \qquad (E > 0). \tag{A2.10b}$$

$$K_3(q) = \int dq \, (E + U \operatorname{sech}^2 \alpha q)^{-\frac{1}{2}}$$

$$= \frac{1}{\alpha \sqrt{-E}} \sin^{-1} \left[\sqrt{\frac{-E}{U + E}} \sinh \alpha q \right] \qquad (-U < E < 0), \tag{A2.11a}$$

$$= \frac{1}{\alpha \sqrt{E}} \sinh^{-1} \left[\sqrt{\frac{E}{U + E}} \sinh \alpha q \right] \qquad (E > 0) \tag{A2.11b}$$

$$K_4(q) = \alpha \int dx \ \frac{x}{\sqrt{Ex^2 + 2Ux - U}} \qquad (q = \alpha x)$$

$$= \alpha \frac{(Ex^2 + 2Ux - U)^{\frac{1}{2}}}{E} + \frac{U\alpha}{(-E)^{\frac{3}{2}}} \sin^{-1} \left(\frac{Ex + U}{\sqrt{U(U+E)}} \right)$$

$$(-U < E < 0) \tag{A2.12a}$$

$$= \alpha \frac{(Ex^2 + 2Ux - U)^{\frac{1}{2}}}{E} - \frac{U\alpha}{E^{\frac{3}{2}}} \cosh^{-1} \left(\frac{Ex + U}{\sqrt{U(U+E)}} \right)$$

$$(E > 0). \tag{A2.12b}$$

INDEX

Printed in the United Kingdom
by Lightning Source UK Ltd.
321